At Sylvan, we believe that everyone can master math skills, and we are glad you have chosen our resources to help your children experience the joy of mathematics as they build crucial reasoning skills. We know that time spent reinforcing lessons learned in school will contribute to understanding and mastery.

Success in math requires more than just memorizing basic facts and algorithms; it also requires children to make connections between the real world and math concepts in order to solve problems. Successful problem solvers will be ready for the challenges of mathematics as they advance to more complex topics and encounter new problems both in school and at home.

We use a research-based, step-by-step process in teaching math at Sylvan that includes thought-provoking math problems and activities. As students increase their success as problem solvers, they become more confident. With increasing confidence, students build even more success. The design of the Sylvan workbooks lays out a roadmap for mathematical learning that is designed to lead your child to success in school.

Included with your purchase of this workbook is a coupon for a discount at a participating Sylvan center. We hope you will use this coupon to further your children's academic journeys. Let us partner with you to support the development of confident, well-prepared, independent learners.

The Sylvan Team

D1418217

Sylvan Learning Center
Unleash your child's potential here

No matter how big or small the academic challenge, every child has the ability to learn. But sometimes children need help making it happen. Sylvan believes every child has the potential to do great things. And, we know better than anyone else how to tap into that academic potential so that a child's future really is full of possibilities. Sylvan Learning Center is the place where your child can build and master the learning skills needed to succeed and unlock the potential you know is there.

The proven, personalized approach of our in-center programs deliver unparalleled results that other supplemental education services simply can't match. Your child's achievements will be seen not only in test scores and report cards but outside the classroom as well. And when he starts achieving his full potential, everyone will know it. You will see a new level of confidence come through in everything he does and every interaction he has.

How can Sylvan's personalized in-center approach help your child unleash his potential?

• Starting with our exclusive Sylvan Skills Assessment®, we pinpoint your child's exact academic needs.

• Then we develop a customized learning plan designed to achieve your child's academic goals.

• Through our method of skill mastery, your child will not only learn and master every skill in his personalized plan, he will be truly motivated and inspired to achieve his full potential.

To get started, included with this Sylvan product purchase is $10 off our exclusive Sylvan Skills Assessment®. Simply use this coupon and contact your local Sylvan Learning Center to set up your appointment.

And to learn more about Sylvan and our innovative in-center programs, call 1-800-EDUCATE or visit www.SylvanLearning.com. *With over 1,000 locations in North America, there is a Sylvan Learning Center near you!*

4th Grade
Math
Games & Puzzles

Published in the United States by Random House, Inc., New York, and in Canada by Random House of Canada Limited, Toronto.

www.tutoring.sylvanlearning.com

Created by Smarterville Productions LLC
Producer & Editorial Direction: The Linguistic Edge
Producer: TJ Trochlil McGreevy
Writer: Amy Kraft
Cover and Interior Illustrations: Shawn Finley and Duendes del Sur
Layout and Art Direction: SunDried Penguin
Director of Product Development: Russell Ginns

First Edition

ISBN: 978-0-375-43043-5

This book is available at special discounts for bulk purchases for sales promotions or premiums. For more information, write to Special Markets/Premium Sales, 1745 Broadway, MD 6-2, New York, New York 10019 or e-mail specialmarkets@randomhouse.com.

PRINTED IN CHINA

10 9 8 7 6 5 4 3 2 1

Contents

Contents

Number Search

9/29/11

WRITE each number. Then CIRCLE it in the puzzle.

HINT: Numbers are across and down only.

1. eighty-four thousand, one hundred sixty-five __84,165__

2. four million, six hundred seventy-two thousand, two hundred forty-four __4,672,244__

3. nine hundred sixty-one thousand, seven hundred twenty-three __961,723__

4. twenty-nine thousand, eight hundred eleven __29,811__

5. one hundred fifteen thousand, seven hundred thirty-six __115,736__

6. two million, eighty-two thousand, six hundred forty-one __2,082,641__

7. five hundred five thousand, six hundred ninety-two __505,692__

8. three million, nine hundred thirty-seven thousand, two hundred sixty __3,937,260__

2	9	8	1	1	0	0	2
9	5	0	5	6	9	2	0
8	1	8	2	4	6	7	8
2	4	4	9	9	1	3	2
0	5	1	1	5	7	3	6
3	4	6	7	2	2	4	4
6	0	5	3	7	3	4	1
3	9	3	7	2	6	0	8

Criss Cross

READ the clues, and WRITE the numbers in the puzzle.

ACROSS

1. two million, five hundred ninety-one thousand, three hundred twenty-four

4. eight million, four hundred sixty-seven thousand, five hundred fifty-three

7. ninety-six thousand, eight hundred twenty-four

10. sixty-four thousand, one hundred ninety-nine

11. one million, one hundred fifty-two thousand, seven hundred three

14. two hundred seventy thousand, three hundred ninety-six

17. six hundred twenty-eight thousand, nine hundred thirty-one

DOWN

2. five hundred seventy-eight thousand, thirty-six

3. three hundred seven thousand, two hundred ninety-four

6. three million, seven hundred forty-two thousand, six hundred eighty

7. nine million, four hundred thirty-one thousand, fifty-two

8. eighty-eight thousand, seven hundred sixteen

9. four million, nine hundred thirteen thousand, five hundred forty-six

12. five million, six hundred nine thousand, four hundred twenty-eight

13. three hundred sixty-four thousand, seven hundred seventy-one

15. seventy-three thousand, nineteen

16. one hundred ninety-seven thousand, three hundred seven

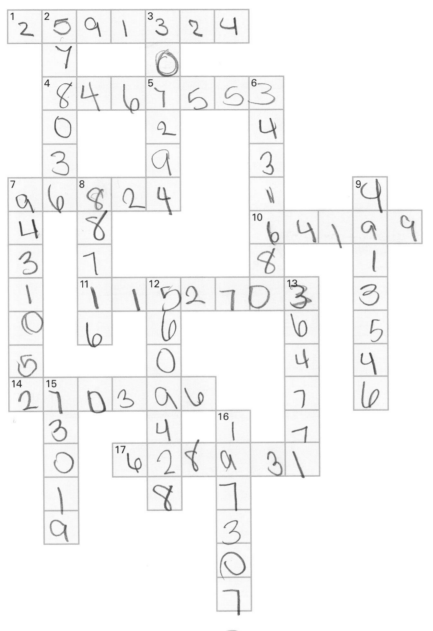

Place Value

Secret Number

10/1/11

DRAW a line to get from the start of the maze to the end without taking any extra paths. WRITE each number you cross in order, starting with the millions place, to find the secret number.

Secret number:

7,428,605

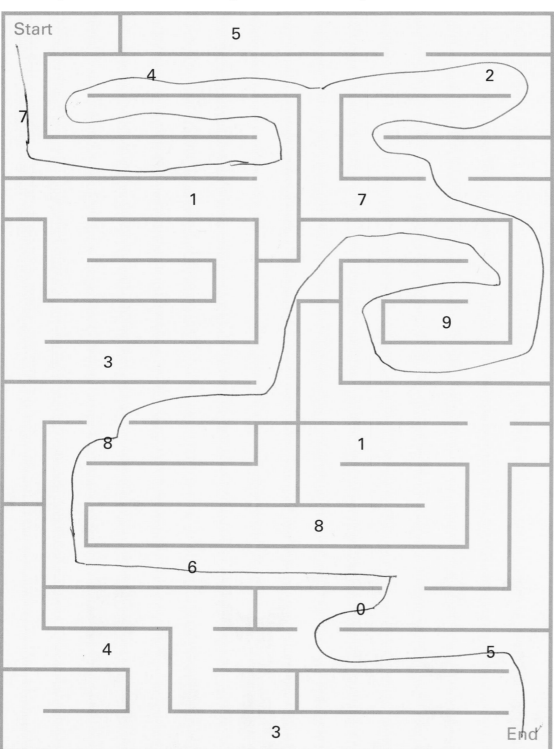

Totally Tangled

10/2/11

Each numbered circle is connected to another numbered circle. FIND the pairs of numbers, and
COLOR any pair that shows a number with that number correctly rounded to the nearest ten thousand.

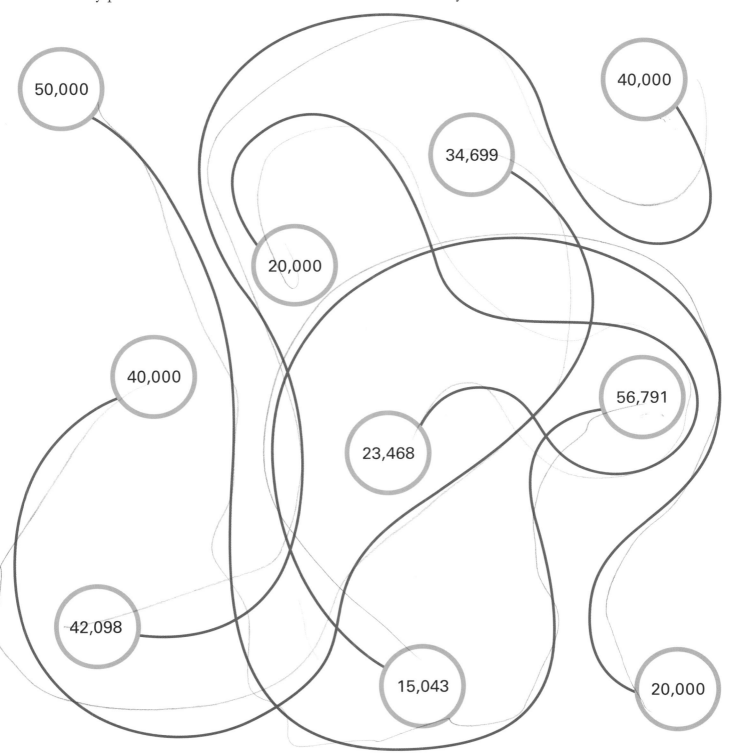

Just Right

10/18/11

WRITE each of the numbers to correctly complete the sentences.

HINT: There may be more than one place to put a number, but you need to use every number.

| 341,156 | 392,382 | 275,319 | 337,236 | 232,981 |
| 384,620 | 228,864 | 382,495 | 246,518 | |

1. _____228,864_____ rounded to the nearest thousand is 229,000.

2. _____341,156_____ rounded to the nearest ten thousand is 340,000.

3. _____ rounded to the nearest hundred thousand is 300,000.

4. _____rounded to the nearest ten thousand is 380,000.

5. _____ rounded to the nearest thousand is 382,000.

6. _____ rounded to the nearest hundred thousand is 400,000.

7. _____ rounded to the nearest ten thousand is 230,000.

8. _____ rounded to the nearest thousand is 337,000.

9. _____ rounded to the nearest hundred thousand is 200,000.

Picking Pairs

10/3/11

DRAW a line to connect each number with that number rounded to the nearest hundred thousand.

Just Right

10/18/11

WRITE each of the numbers to correctly complete the sentence.

HINT: There may be more than one place to put a number, but you need to use every number.

5,418,163	5,908,752	5,826,138	6,692,556	5,237,564
6,694,204	5,879,215	5,418,921	6,563,827	

1. _____ rounded to the nearest million is 5,000,000.

2. _____ rounded to the nearest hundred thousand is 5,400,000.

3. _____ rounded to the nearest ten thousand is 5,910,000.

4. _____ rounded to the nearest hundred thousand is 6,700,000.

5. _____ rounded to the nearest thousand is 6,694,000.

6. _____ rounded to the nearest million is 7,000,000.

7. _____ rounded to the nearest hundred thousand is 5,900,000.

8. _____ rounded to the nearest thousand is 5,419,000.

9. _____ rounded to the nearest million is 6,000,000.

Number Factory

10/5/11

WRITE the numbers that will come out of each machine.

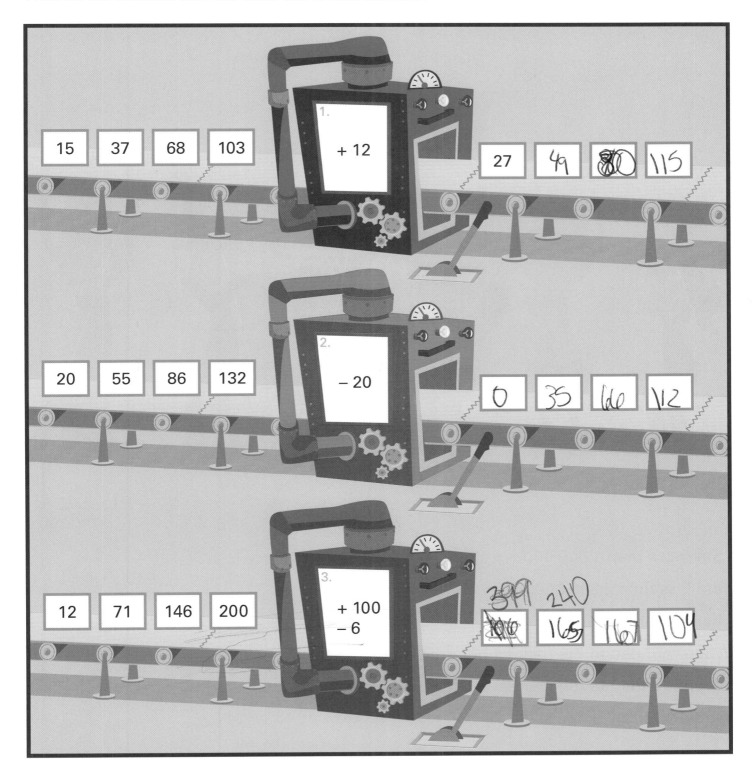

1. + 12

15 37 68 103 → 27 49 80 115

2. − 20

20 55 86 132 → 0 35 66 112

3. + 100 − 6

12 71 146 200 → 106 165 167 104

Who Am I?

10/6/11

READ the clues, and CIRCLE the mystery number.

HINT: Cross out any number that does not match the clues.

I am more than 1,500,000.

I am less than 2,500,000.

I have a 1 in the millions place.

When rounded to the nearest hundred thousand, I'm 1,800,000.

When rounded to the nearest ten thousand, I'm 1,840,000.

Who am I?

Number Factory

10/7/11

WRITE the numbers that will come out of each machine.

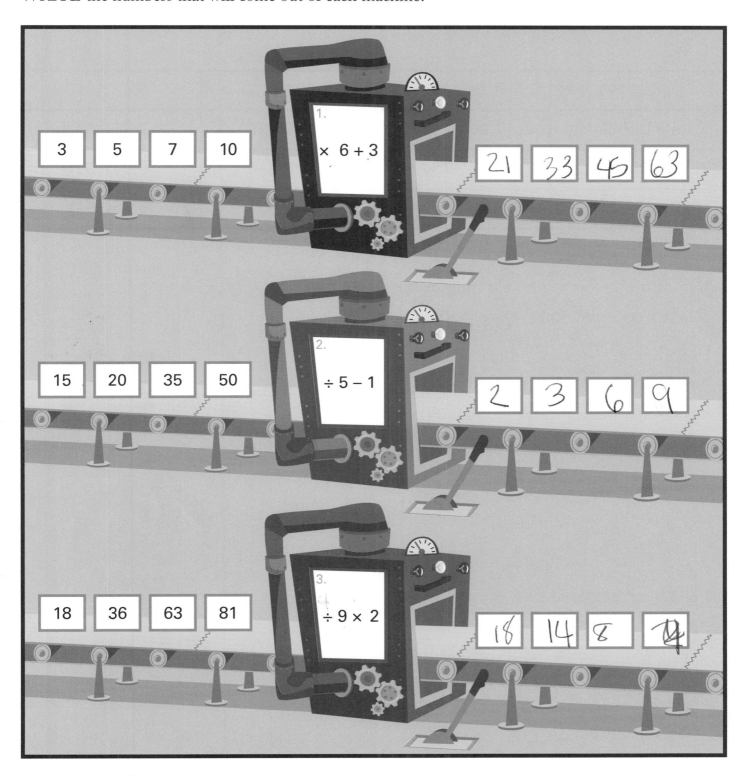

1. × 6 + 3

| 3 | 5 | 7 | 10 |

| 21 | 33 | 45 | 63 |

2. ÷ 5 − 1

| 15 | 20 | 35 | 50 |

| 2 | 3 | 6 | 9 |

3. ÷ 9 × 2

| 18 | 36 | 63 | 81 |

| 18 | 14 | 8 | 4 |

Last One Standing

10/4/11

CROSS OUT each number that does not match the clues until one number is left.

HINT: Follow the clues in order. The last number left should match all of the clues.

7,235,192	6,850,561	7,091,535	6,431,830
6,818,567	7,139,408	6,328,320	7,557,241
7,574,688	8,002,152	7,978,614	6,929,328
6,513,216	6,489,773	7,156,902	6,830,515

When rounded to the nearest million, it is 7,000,000.

It is less than 7,100,000.

It is more than 6,700,000.

It has a 5 in the hundreds place.

It has an 8 in the hundred thousands place.

When rounded to the nearest ten thousand, it is 6,820,000.

Number Factory

WRITE the numbers that belong on the side of each machine.

HINT: The numbers are all between 1 and 10.

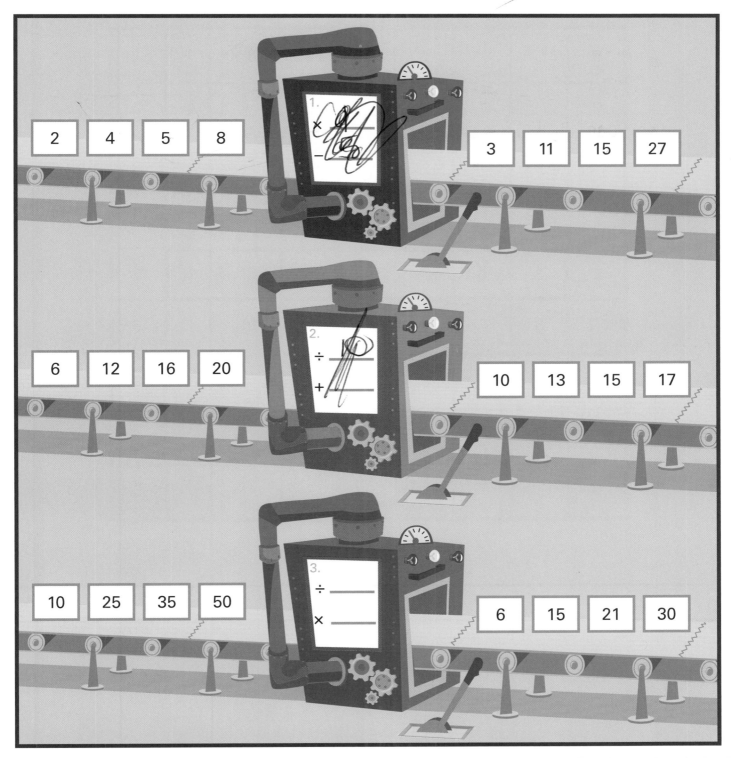

Super Sudoku

10/8/11

WRITE the numbers 1 through 9 so that each row, column, and box has all nine numbers. Then WRITE the number words for the number outlined in red.

9	5	3	1	4	6	8		2
2	7	1	5	8	9		4	3
4	6	8	7	2	3	5		
	3		2	6			5	4
1	2	6			5		8	7
	8		3	1		9		6
6		5	8				3	9
	9		6	5	4	7		
8		7			2	4		5

_____

_____

Adding

4

Pipe Down

10/9/11

WRITE the missing number. Then FOLLOW the pipe, and WRITE the same number in the
next problem.

3939
− 1506
2433

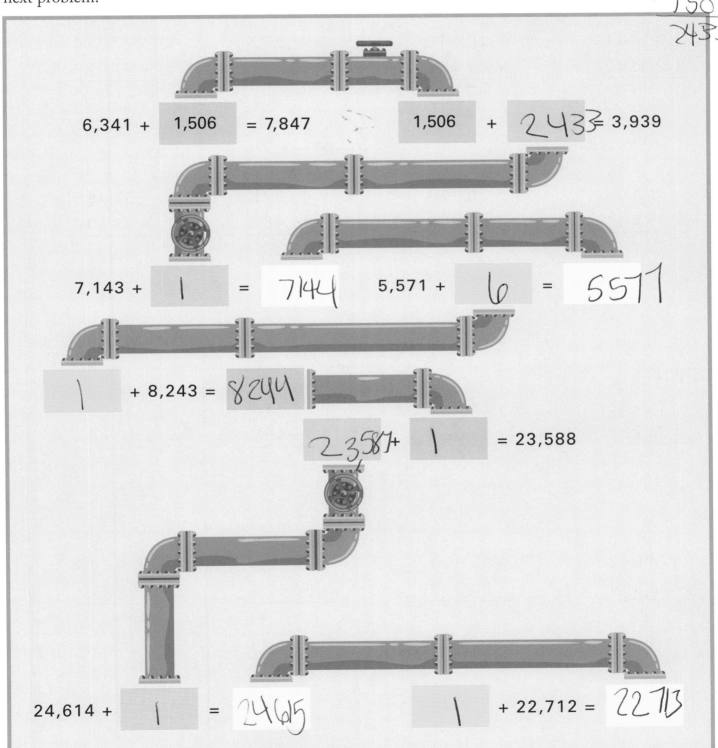

6,341 + 1,506 = 7,847   1,506 + 2433 = 3,939

7,143 + 1 = 7144   5,571 + 6 = 5577

1 + 8,243 = 8244

23587 + 1 = 23,588

24,614 + 1 = 24615   1 + 22,712 = 22713

Adding

Criss Cross

SOLVE the addition problems, and WRITE the sums in the puzzle.

ACROSS

1.
```
  36,981
+ 28,302
```
65,283

3.
```
  20,046
+ 17,060
```
37,106

5.
```
  41,989
+ 16,138
```
58127

7.
```
  27,724
+ 21,628
```
49302

9.
```
  42,105
+ 38,541
```
80646

10.
```
  26,028
+ 44,164
```
70192

12.
```
  43,778
+ 14,464
```
58232

13.
```
  52,995
+ 31,736
```
84731

DOWN

1.
```
  36,268
+ 33,286
```
69554

2.
```
  68,692
+ 13,043
```
81,735

4.
```
  39,614
+ 25,460
```
65074

6.
```
  17,867
+ 10,165
```
28032

8.
```
  74,943
+ 23,987
```
98,930

9.
```
  52,030
+ 30,938
```
82968

10.
```
  36,439
+ 35,675
```
72114

11.
```
  29,625
+ 22,812
```
52,437

Adding

Number Search

WRITE each sum. Then CIRCLE it in the puzzle.

HINT: Numbers are across and down only.

10/17/11

1. 48,350
    + 28,627
    76,977

2. 16,129
    + 69,414
    85,543

3. 39,524
    + 11,825
    51,349

4. 36,942
    + 22,926
    59,868

5. 85,924
    + 13,834
    99,758

6. 46,561
    + 15,811
    62,372

7. 21,842
    + 18,861
    40,703

8. 43,527
    + 27,703
    71,230

4	0	7	0	3	6	5	7
1	3	9	8	5	2	2	6
0	9	9	5	1	3	4	9
8	6	7	0	4	7	9	7
5	4	5	7	9	2	0	7
5	9	8	6	8	3	1	5
4	9	2	7	1	2	3	0
3	1	1	4	0	8	7	4

Pipe Down

10/19/11

WRITE the missing number. Then FOLLOW the pipe, and WRITE the same number in the next problem.

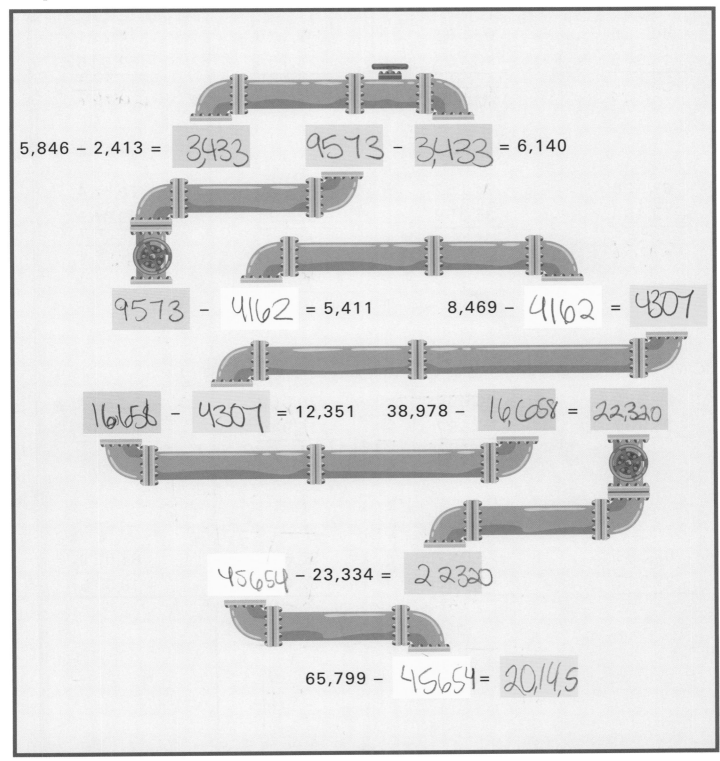

5,846 − 2,413 = 3,433   9573 − 3,433 = 6,140

9573 − 4162 = 5,411   8,469 − 4162 = 4307

16165$ − 4307 = 12,351   38,978 − 16,658 = 22,320

45654 − 23,334 = 22320

65,799 − 45654 = 20,145

21

Criss Cross

10/19/11 →

SOLVE the subtraction problems, and WRITE the differences in the puzzle.

ACROSS

1.
$$
\begin{array}{r}
50,336 \\
- 14,845 \\
\hline
35491
\end{array}
$$

5.
$$
\begin{array}{r}
98,565 \\
- 18,133 \\
\hline
80432
\end{array}
$$

7.
$$
\begin{array}{r}
48,130 \\
- 24,154 \\
\hline
23976
\end{array}
$$

9.
$$
\begin{array}{r}
74,221 \\
- 52,516 \\
\hline
21705
\end{array}
$$

10.
$$
\begin{array}{r}
63,920 \\
- 12,462 \\
\hline
51458
\end{array}
$$

11.
$$
\begin{array}{r}
84,162 \\
- 35,059 \\
\hline
49,103
\end{array}
$$

15.
$$
\begin{array}{r}
86,920 \\
- 62,591 \\
\hline
24,329
\end{array}
$$

16.
$$
\begin{array}{r}
70,644 \\
- 43,969 \\
\hline
26675
\end{array}
$$

DOWN

2.
$$
\begin{array}{r}
89,299 \\
- 30,887 \\
\hline
58,412
\end{array}
$$

3.
$$
\begin{array}{r}
26,604 \\
- 10,597 \\
\hline
16007
\end{array}
$$

4.
$$
\begin{array}{r}
79,867 \\
- 16,943 \\
\hline
62924
\end{array}
$$

6.
$$
\begin{array}{r}
43,837 \\
- 25,911 \\
\hline
17326
\end{array}
$$

8.
$$
\begin{array}{r}
63,391 \\
- 27,917 \\
\hline
35474
\end{array}
$$

12.
$$
\begin{array}{r}
76,012 \\
- 63,060 \\
\hline
12,952
\end{array}
$$

13.
$$
\begin{array}{r}
90,653 \\
- 52,437 \\
\hline
38,215
\end{array}
$$

14.
$$
\begin{array}{r}
81,033 \\
- 18,815 \\
\hline
62218
\end{array}
$$

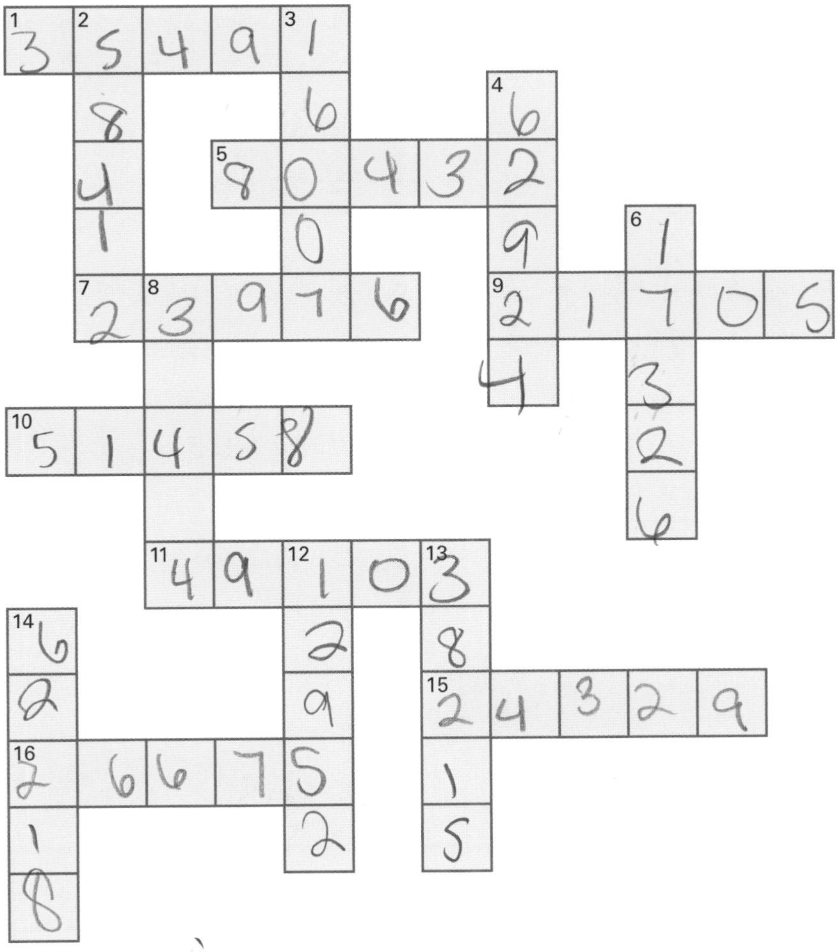
1 3 2 5 4 9 3 1
2 5 8 4 1 2
3 1 6 0 0 9
4 6 2 9 2 4
5 8 0 4 3 2
6 1 7 3 2 6
7 2 3 9 7 6
9 2 1 7 0 5
10 5 1 4 5 8
11 4 9 1 0 3
12 1 2 9 5 2
13 3 8 2 1 5
14 6 2 2 1 8
15 2 4 3 2 9
16 2 6 6 7 5

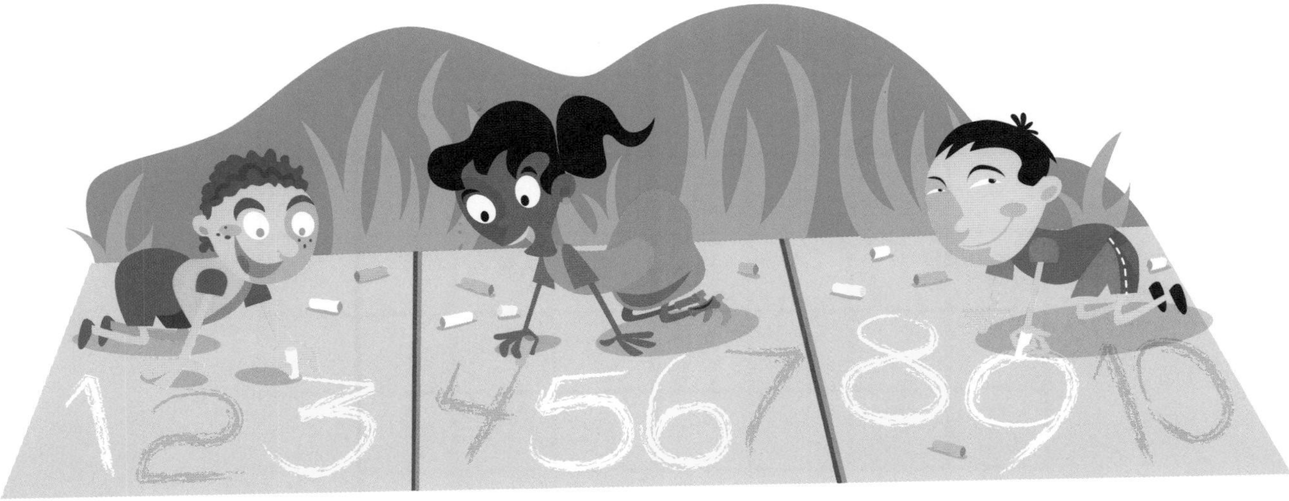
1 2 3 4 5 6 7 8 9 10

Number Search

✓ 10/20/11

WRITE each difference. Then CIRCLE it in the puzzle.

HINT: Numbers are across and down only.

1.
```
    4 9 12 10 18
    5 0,3 0 8
  - 2 7,9 9 7
    2 2,3 1 1
```

2.
```
    8 14 8 10
    9 4,6 9 0
  - 1 8,4 7 7
      7 6,2 1 3
```

3.
```
    7 13 7 11
    8 3,8 1 8
  - 2 8,5 9 4
      5 5 2 2 4
```

4.
```
    8 10   1 10
    9 0,6 2 0
  - 1 9,6 0 2
    7 1,0 1 8
```

5.
```
    5 18 11 15
    6 9,2 6 2 12
  - 1 9,3 8 6
    4 9,8 7 6
```

6.
```
    4 14 4 12 12
    5 4,5 3 2
  - 3 6,2 9 6
    1 8,2 3 6
```

7.
```
          7 14
    8 7,8 5 5 15
  - 2 6,7 6 9
      6 1,0 8 6
```

8.
```
    4 11
    4 5,1 9 9
  - 1 0,2 2 7
    3 4,9 7 2
```

1	8	2	3	6	9	6	1
7	6	2	4	6	0	1	7
1	1	0	2	3	5	0	1
3	5	5	2	2	4	8	0
4	2	2	3	5	9	6	1
9	0	4	1	6	8	1	8
7	6	2	1	3	7	6	2
2	9	8	8	3	6	7	2

Picking Pairs

10/20/11 ✓

ESTIMATE each sum or difference by rounding to the nearest ten thousand. DRAW a line to connect each problem with the correct estimate of the sum or difference.

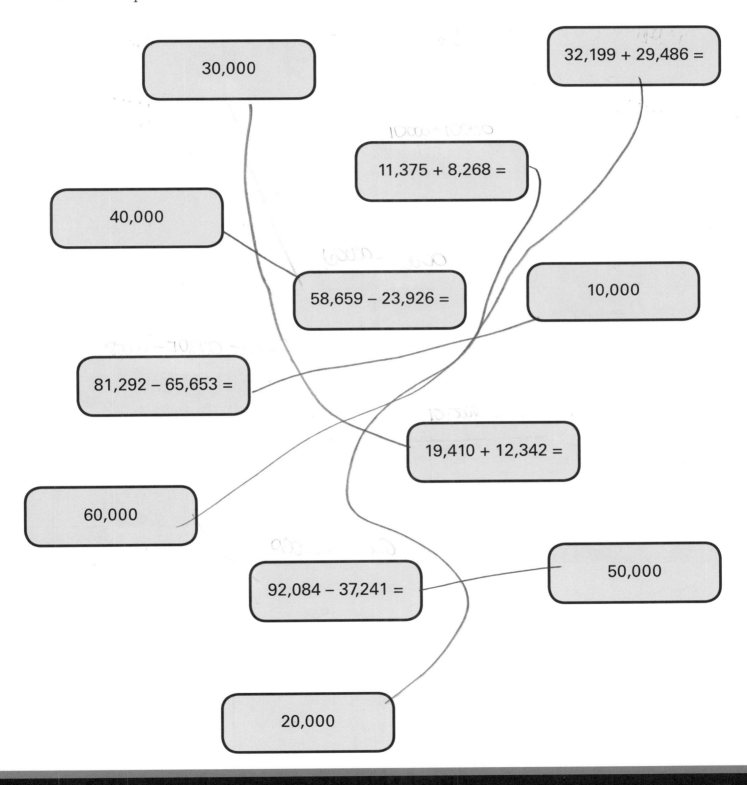

30,000

32,199 + 29,486 =

11,375 + 8,268 =

40,000

58,659 − 23,926 =

10,000

81,292 − 65,653 =

19,410 + 12,342 =

60,000

50,000

92,084 − 37,241 =

20,000

Hidden Design

ESTIMATE each sum or difference by rounding each number to the nearest thousand.
Then COLOR the squares that match the numbers to see the hidden design.

23,671 + 5,092 =

68,349 – 42,688 =

12,765 + 18,135 =

50,913 – 16,320 =

14,826 + 22,751 =

40,559 – 13,429 =

25,000	38,000	25,000	28,000	29,000	31,000	29,000	28,000
38,000		38,000	25,000	28,000	29,000	28,000	25,000
	31,000		38,000	25,000	28,000	25,000	38,000
31,000	29,000	31,000		38,000	25,000	38,000	
29,000	28,000	29,000	31,000		38,000		31,000
28,000	25,000	28,000	29,000	31,000		31,000	29,000
25,000	38,000	25,000	28,000	29,000	31,000	29,000	28,000
38,000		38,000	25,000	28,000	29,000	28,000	25,000

10/21

Pipe Down

WRITE the missing number. Then FOLLOW the pipe, and WRITE the same number in the next problem.

$15,439 + 44,748 =$ 60187   60187 $-$ 25633 $= 34,554$

25633 $+$ 38502 $= 64,135$ $82,146 -$ 38502 $=$ 43644

50559 $+$ 43644 $= 94,203$ 50559 $- 14,618 =$ 35941

17692 $+ 18,249 =$ 35941

$85,126 -$ 17692 $=$ 67434

Super Square

10/21/11

WRITE numbers in the empty squares to finish all of the subtraction problems.

85,000	–	53,218	=	31,782
–		–		–
72,883	–	46,914	=	25,969
=		=		=
12,117	–	6,304	=	5,813

Code Breaker

SOLVE each problem. WRITE the letter that matches each product to solve the riddle.

6 × 5	8 × 2	9 × 4	5 × 5	3 × 8	1 0 × 6
1 **30**	2 **16**	3 **36**	4 **25**	5 **24**	6 **60**
M	R	W	U	V	H

7 × 1	9 × 8	6 × 8	2 × 9	9 × 5	1 0 × 4
7 **7**	8 **72**	9 **48**	11 **18**	12 **45**	13 **40**
Y	E	P	T	O	I

Where do you find a turtle with no legs?

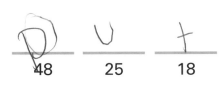

W	h	e	r	e	v	e	r
36	60	72	16	72	24	72	16

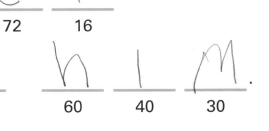

Y	O	U	p	u	t	h	I	M
7	45	25	48	25	18	60	40	30

Multiplication Facts

Gridlock

WRITE numbers so that the product of the rows and columns is correct.

HINT: Use only the numbers 1 through 10.

Example:

	3	5
4	12	20
8	24	40

$3 \times 4 = 12$

$3 \times 8 = 24$

$5 \times 4 = 20$

$5 \times 8 = 40$

Work Box

```
 3  4  5  6      1
 6  8  10 12              12        12
 9  12 15 18         x  6      x 11
 12 16 20 24          72        12
 15 20 24 30                  +120
 18 24 30 36          1       132
 21 28 35 42         12
 24 32 40 48        x  7       12
 27 36 45 54        84       x 10
 40 50 60                    00
                             120
```

	4	7
1	4	7
2	8	14

	5	6
1	5	6
3	15	18

	2	7
9	18	63
10	20	70

	10	16
1	10	16
3	30	48

	3	8
7	21	56
9	27	72

	5	6
7	35	42
9	45	54

Pipe Down

WRITE the missing number. Then FOLLOW the pipe, and WRITE the same number in the next problem.

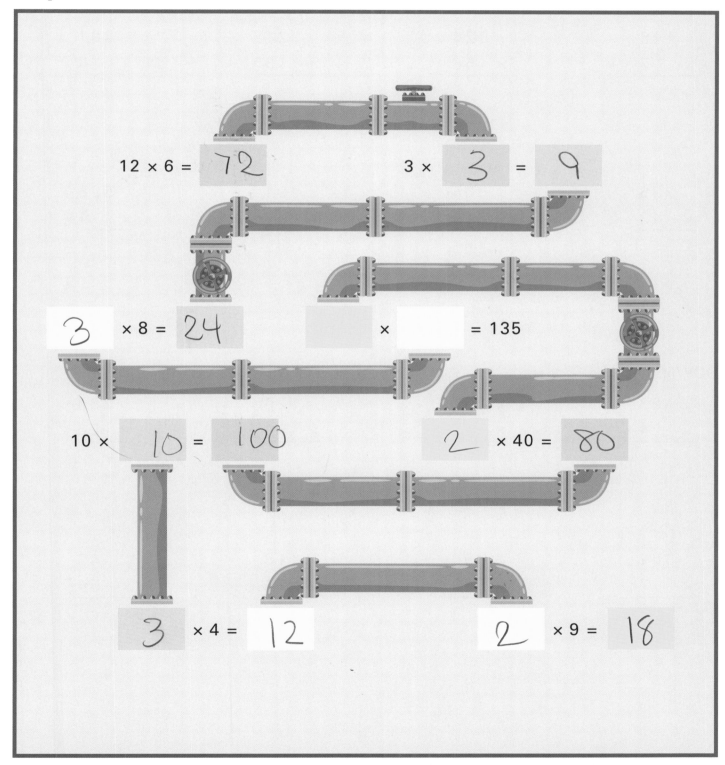

12 × 6 = 72

3 × 3 = 9

3 × 8 = 24

× = 135

10 × 10 = 100

2 × 40 = 80

3 × 4 = 12

2 × 9 = 18

10/22

Criss Cross

SOLVE the multiplication problems, and WRITE the products in the puzzle.

ACROSS

1.
```
  148
×   7
```

4.
```
  824
×  68
```

5.
```
   21
×  13
```

7.
```
  401
×   8
```

9.
```
  335
×  43
```

10.
```
   64
×  45
```

11.
```
  491
×  52
```

12.
```
  528
×  71
```

DOWN

1.
```
   53
×  24
```

2.
```
  137
×   5
```

3.
```
  120
×  34
```

6.
```
  289
×  26
```

7.
```
  501
×   7
```

8.
```
  896
×  93
```

9.
```
  118
×  14
```

10.
```
   42
×  54
```

10/22

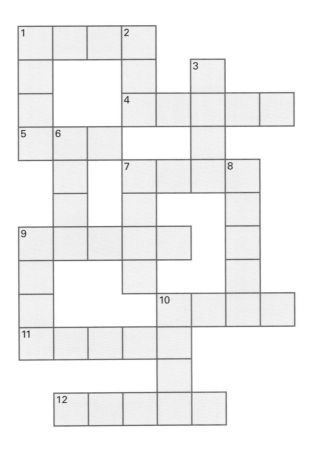

Super Square

WRITE numbers in the empty squares to finish all of the multiplication problems.

13	×	11	=	
×		×		×
6	×		=	
=		=		=
	×	154	=	

Code Breaker

SOLVE each problem. WRITE the letter that matches each quotient to solve the riddle.

1 $\overset{7}{9\overline{)63}}$	2 $\overset{8}{7\overline{)56}}$	3 $\overset{10}{3\overline{)30}}$	4 $\overset{1}{12\overline{)12}}$	5 $\overset{4}{4\overline{)16}}$	6 $\overset{2}{9\overline{)18}}$
C	O	H	A	T	E

7 $\overset{5}{9\overline{)45}}$	8 $\overset{11}{2\overline{)22}}$	9 $\overset{3}{8\overline{)24}}$	10 $\overset{12}{3\overline{)36}}$	11 $\overset{6}{5\overline{)30}}$	12 $\overset{9}{8\overline{)72}}$
G	F	U	I	L	S

How did the frog make the baseball team?

$\underset{10}{H} \quad \underset{2}{E}$

$\underset{7}{C} \quad \underset{1}{A} \quad \underset{3}{U} \quad \underset{5}{g} \quad \underset{10}{h} \quad \underset{4}{t}$

$\underset{1}{A} \qquad \underset{6}{L} \quad \underset{8}{O} \quad \underset{4}{T}$

$\underset{8}{O} \quad \underset{11}{f} \qquad \underset{11}{f} \quad \underset{6}{l} \quad \underset{12}{i} \quad \underset{2}{e} \quad \underset{9}{S}.$

Number Factory

10/24

WRITE the numbers that will come out of each machine.

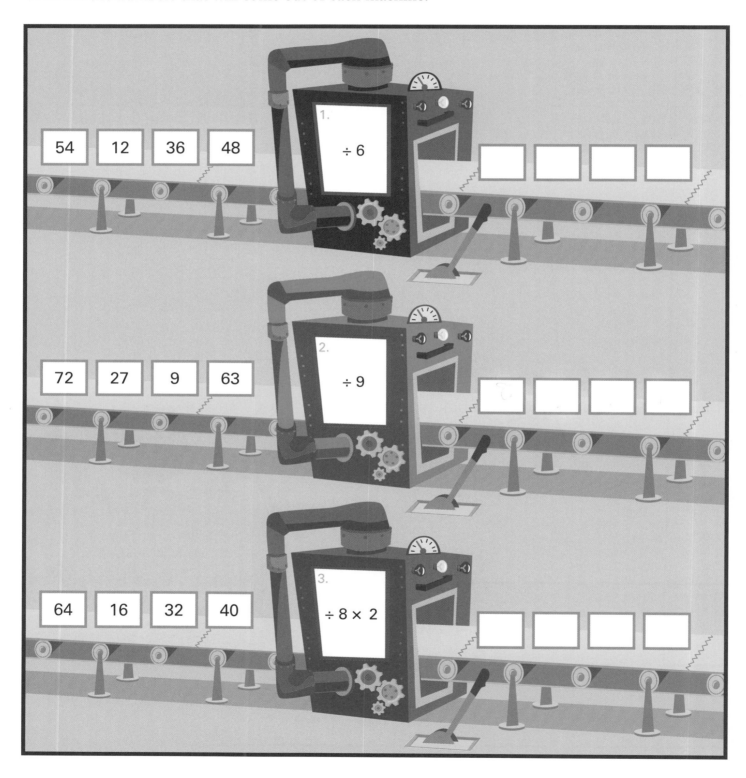

10/24

Pipe Down

WRITE the missing number. Then FOLLOW the pipe, and WRITE the same number in the next problem.

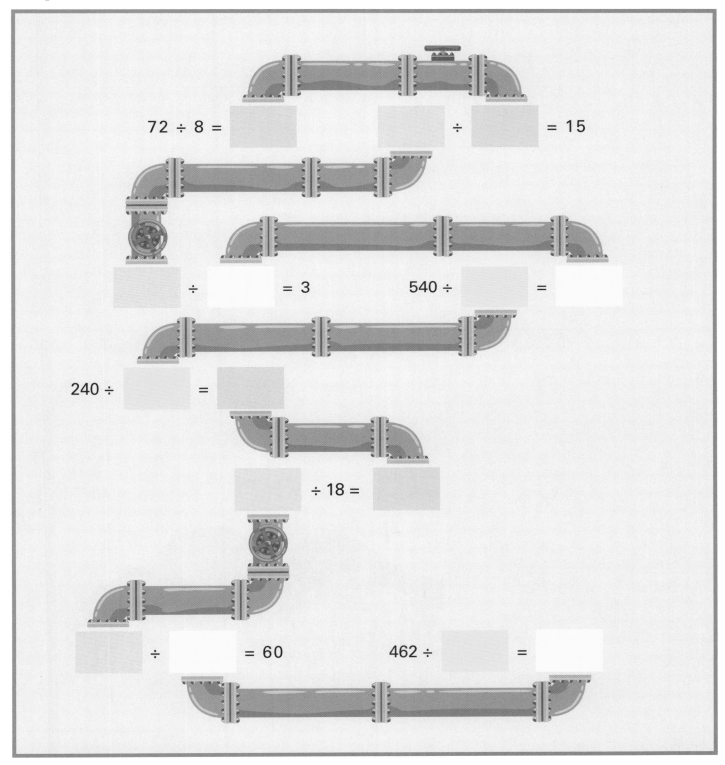

$72 \div 8 = \boxed{}$ $\boxed{} \div \boxed{} = 15$

$\boxed{} \div \boxed{} = 3$ $540 \div \boxed{} = \boxed{}$

$240 \div \boxed{} = \boxed{}$

$\boxed{} \div 18 = \boxed{}$

$\boxed{} \div \boxed{} = 60$ $462 \div \boxed{} = \boxed{}$

Criss Cross

SOLVE the division problems, and WRITE the quotients in the puzzle.

ACROSS

1	4	7	8	9	11

7)924   2)836   2)682   8)840   35)735   58)696

13	14	15	16	19	21

3)759   2)920   17)510   28)812   3)873   2)770

24	25

7)840   4)972

DOWN

1	2	3	4
42)546	8)272	3)630	2)838

5	6	9	10	11	12
46)460	11)935	4)892	5)750	6)972	3)627

17	18	19	20	22	23
2)620	1)532	31)651	8)736	11)924	16)848

Super Square

WRITE numbers in the empty squares to finish all of the division problems.

972	÷	36	=	
÷		÷		÷
54	÷		=	
=		=		=
	÷	6	=	

Crossing Paths

WRITE the missing numbers.

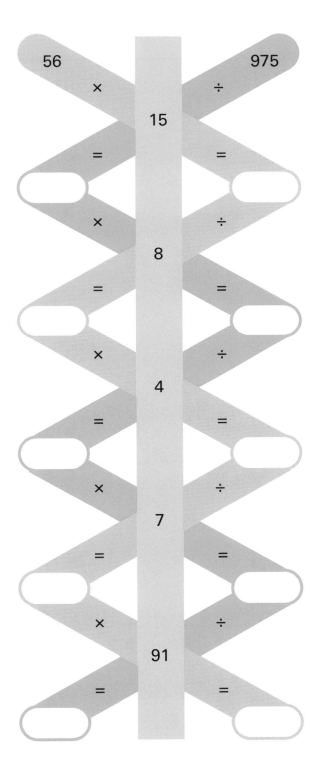

Left puzzle:

Left	Center	Right
24		112
×		÷
	8	
=		=
___		___
×	÷	
	16	
=		=
___		___
×	÷	
	28	
=		=
___		___
×	÷	
	84	
=		=
___		___
×	÷	
	32	
=		=
___		___

Right puzzle:

Left	Center	Right
56		975
×		÷
	15	
=		=
___		___
×	÷	
	8	
=		=
___		___
×	÷	
	4	
=		=
___		___
×	÷	
	7	
=		=
___		___
×	÷	
	91	
=		=
___		___

Number Factory

WRITE the numbers that will come out of each machine.

HINT: Do the operations on each machine in order from top to bottom.

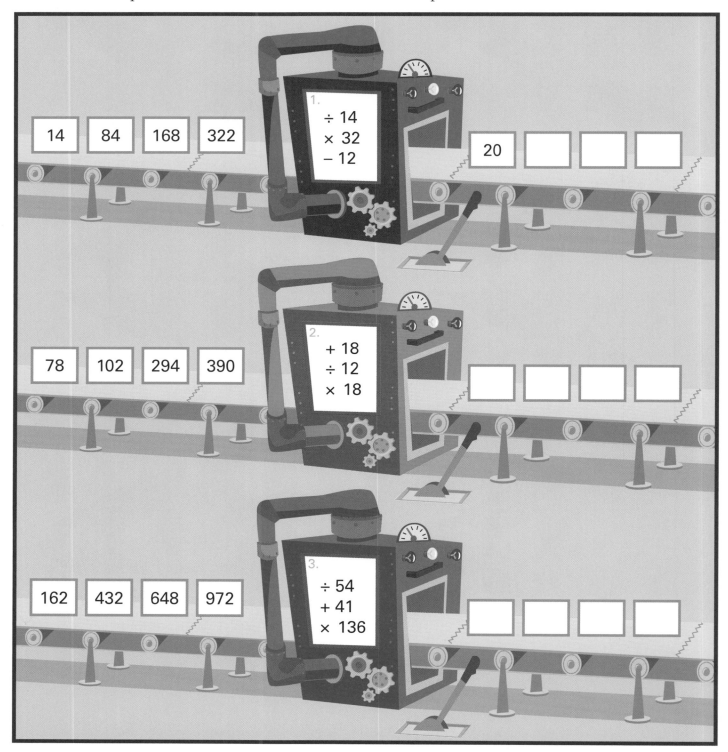

1.
÷ 14
× 32
− 12

14   84   168   322

20

2.
+ 18
÷ 12
× 18

78   102   294   390

3.
÷ 54
+ 41
× 136

162   432   648   972

What's the Password?

WRITE the letters that form a fraction of each word. Then WRITE the letters in order to find the secret password.

1. The first $\frac{1}{3}$ of **SURVEY** _____ SU

2. The first $\frac{1}{7}$ of **MISSING** _____ M

3. The first $\frac{2}{9}$ of **MESMERIZE** _____ M

4. The last $\frac{1}{6}$ of **WINTER** _____ ER

5. The first $\frac{3}{7}$ of **VACCINE** _____ VA

6. The middle $\frac{1}{5}$ of **GRAVY** _____ A

7. The first $\frac{1}{2}$ of **TINY** _____ TI

8. The last $\frac{2}{7}$ of **HEXAGON** _____ on

Password:

Su mmer
S u mmer
Vacation

Picking Pairs

DRAW a line to connect each decimal with the correct picture.

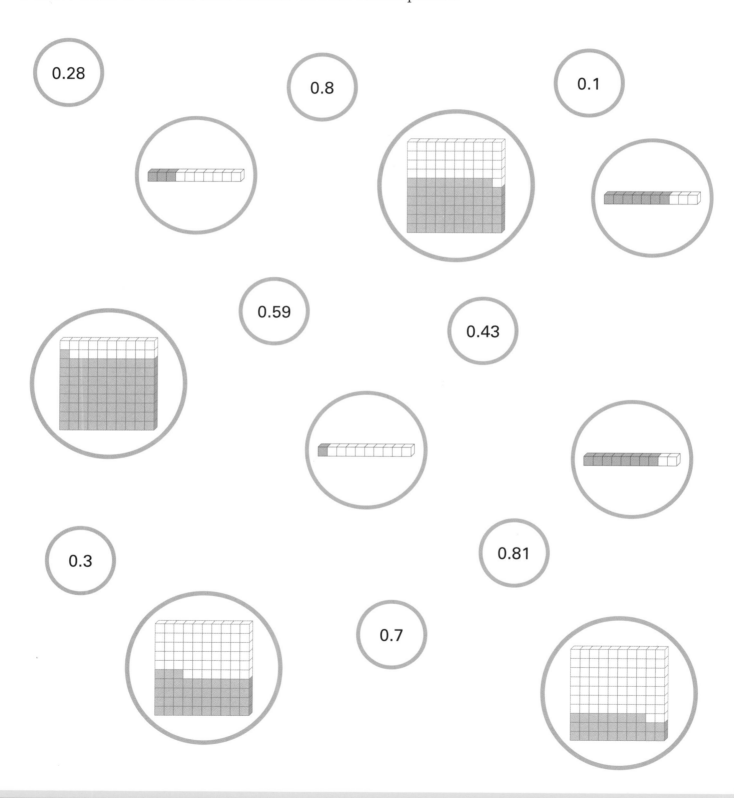

High Card

Use the cards from page 111. READ the rules. PLAY the game!

Rules: Two players

1. Decide if you want to play with fractions or decimals, and face the selected side down.
2. Deal all of the cards so both players have an equal stack of cards.
3. Both players flip a card over at the same time.
4. Whoever has the card with the biggest fraction or decimal gets to keep the cards.

The player with the most cards at the end wins!

Examples:

$\frac{5}{8}$ is bigger than $\frac{1}{6}$

0.73 is bigger than 0.4

Totally Tangled

Each circle is connected to another numbered circle. FIND the pairs of decimals, and COLOR the smaller decimal.

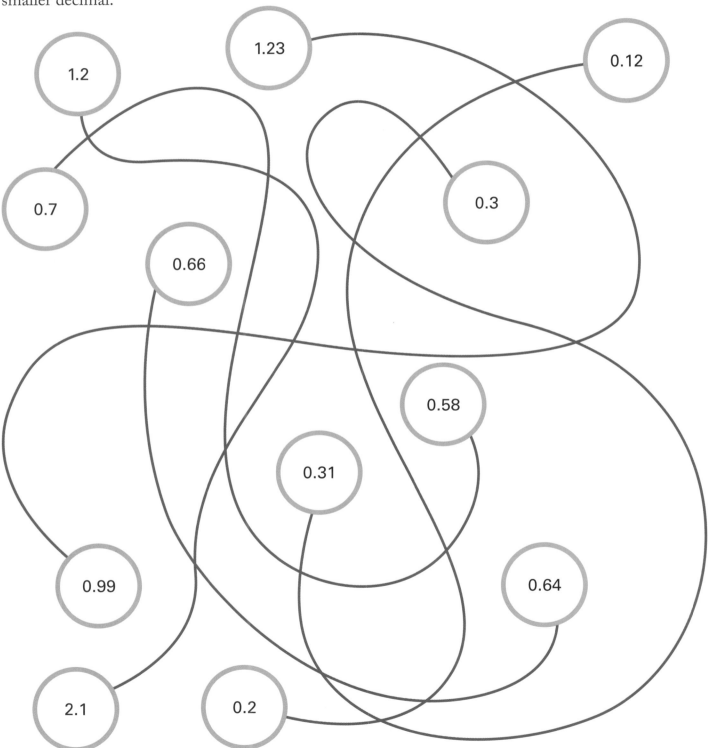

Code Breaker

SOLVE each problem. WRITE the letter that matches each sum to solve the riddle.

$\dfrac{1}{4} + \dfrac{2}{4} =$ ___ [1] **H**	$\dfrac{1}{6} + \dfrac{4}{6} =$ ___ [2] **B**	$\dfrac{2}{3} + \dfrac{1}{3} =$ ___ [3] **T**
$\dfrac{1}{9} + \dfrac{4}{9} =$ ___ [4] **L**	$\dfrac{3}{8} + \dfrac{4}{8} =$ ___ [5] **E**	$\dfrac{1}{5} + \dfrac{1}{5} =$ ___ [6] **R**
$\dfrac{1}{3} + \dfrac{1}{3} =$ ___ [7] **A**	$\dfrac{3}{7} + \dfrac{2}{7} =$ ___ [8] **I**	$\dfrac{2}{8} + \dfrac{3}{8} =$ ___ [9] **Y**

Which building has the most stories?

___ ___ ___
$\dfrac{3}{3}$ $\dfrac{3}{4}$ $\dfrac{7}{8}$

___ ___ ___ ___ ___ ___ ___ .
$\dfrac{5}{9}$ $\dfrac{5}{7}$ $\dfrac{5}{6}$ $\dfrac{2}{5}$ $\dfrac{2}{3}$ $\dfrac{2}{5}$ $\dfrac{5}{8}$

Number Factory

WRITE the fractions that will come out of each machine.

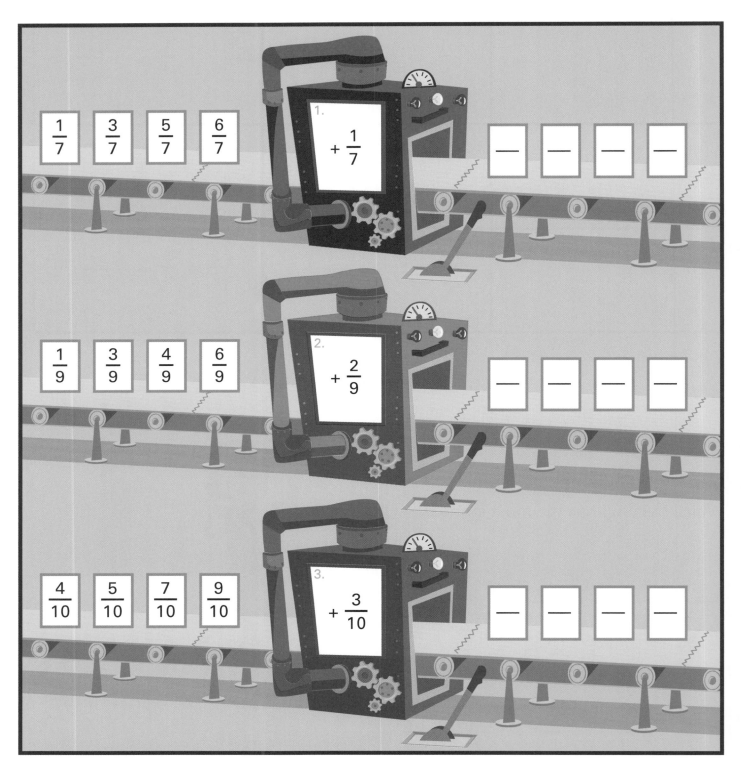

Code Breaker

SOLVE each problem. WRITE the letter that matches each difference to solve the riddle.

$\frac{3}{5} - \frac{1}{5} =$ ___ (1)	$\frac{9}{12} - \frac{2}{12} =$ ___ (2)	$\frac{7}{8} - \frac{4}{8} =$ ___ (3)	$\frac{3}{3} - \frac{2}{3} =$ ___ (4)
E	T	U	R
$\frac{4}{6} - \frac{1}{6} =$ ___ (5)	$\frac{7}{7} - \frac{6}{7} =$ ___ (6)	$\frac{3}{4} - \frac{2}{4} =$ ___ (7)	$\frac{4}{5} - \frac{1}{5} =$ ___ (8)
S	P	H	Y
$\frac{8}{9} - \frac{6}{9} =$ ___ (9)	$\frac{6}{11} - \frac{4}{11} =$ ___ (10)	$\frac{6}{6} - \frac{1}{6} =$ ___ (11)	$\frac{8}{10} - \frac{5}{10} =$ ___ (12)
N	I	L	O

How do snails talk to each other?

___ ___ ___ ___   ___ ___ ___
$\frac{7}{12}$ $\frac{1}{4}$ $\frac{2}{5}$ $\frac{3}{5}$ $\frac{3}{8}$ $\frac{3}{6}$ $\frac{2}{5}$

___ ___ ___ ___ ___   ___ ___ ___ ___ ___
$\frac{7}{12}$ $\frac{1}{4}$ $\frac{2}{5}$ $\frac{2}{11}$ $\frac{1}{3}$ $\frac{3}{6}$ $\frac{1}{4}$ $\frac{2}{5}$ $\frac{5}{6}$ $\frac{5}{6}$

___ ___ ___ ___ ___ ___ .
$\frac{1}{7}$ $\frac{1}{4}$ $\frac{3}{10}$ $\frac{2}{9}$ $\frac{2}{5}$ $\frac{3}{6}$

Number Factory

WRITE the fractions that will come out of each machine.

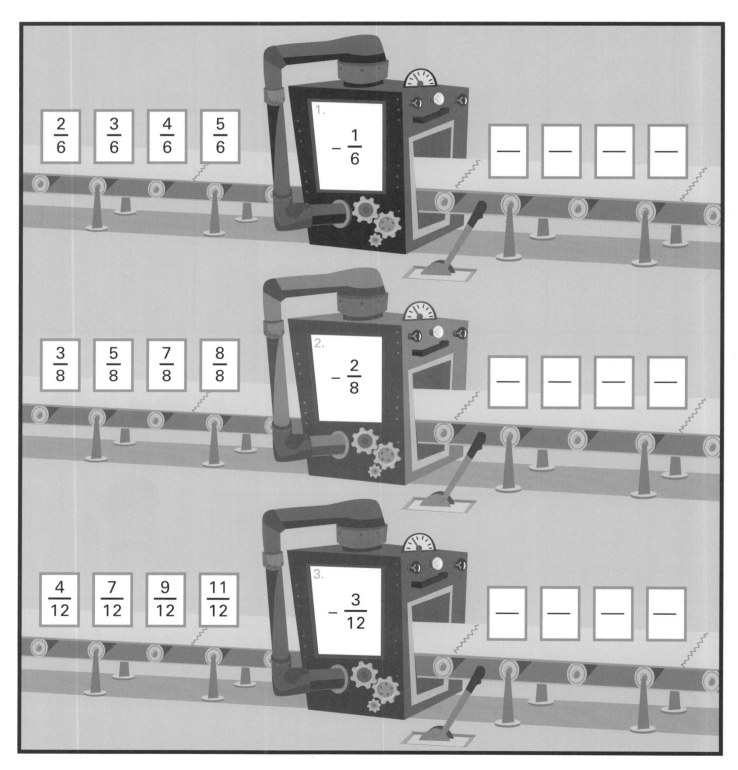

Code Breaker

SOLVE each problem. WRITE the letter that matches each sum or difference to solve the riddle.

```
   3.5          1.4          2.38         4.92         3.82          5.6
 + 6.2        + 2.1        + 1.43       + 3.55       + 4.2         + 4.59
```
1 9.7 ✓   2 3.5 ✓   3 3.81 ✓   4 8.47   5 8.02   6 10.19

I   T   E   A   N   L

```
   8.6          9.9          7.05        10.51        12.43
 - 2.5        - 4.3        - 2.62       - 8.67       - 9.5
```
7 6.1   8 5.6   9 3.43   10 1.84   11 2.93

W   O   M   G   B

What time is it when an elephant sits on your table?

T I M E _ T _ _ E T
3.5 9.7 4.43 3.81   3.5 5.6   1.84 3.81 3.5

A _ N E W _ T A B L E .
8.47   8.02 3.81 6.1   3.5 8.47 2.93 10.19 3.81

Number Factory

WRITE the numbers that will come out of each machine.

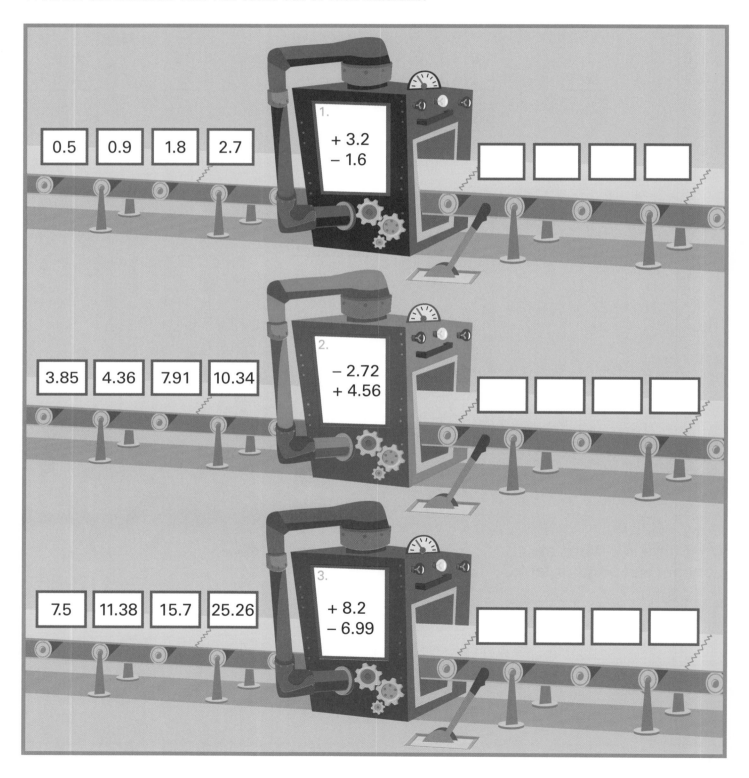

Pipe Down

WRITE the missing number. Then FOLLOW the pipe, and WRITE the same number in the next problem.

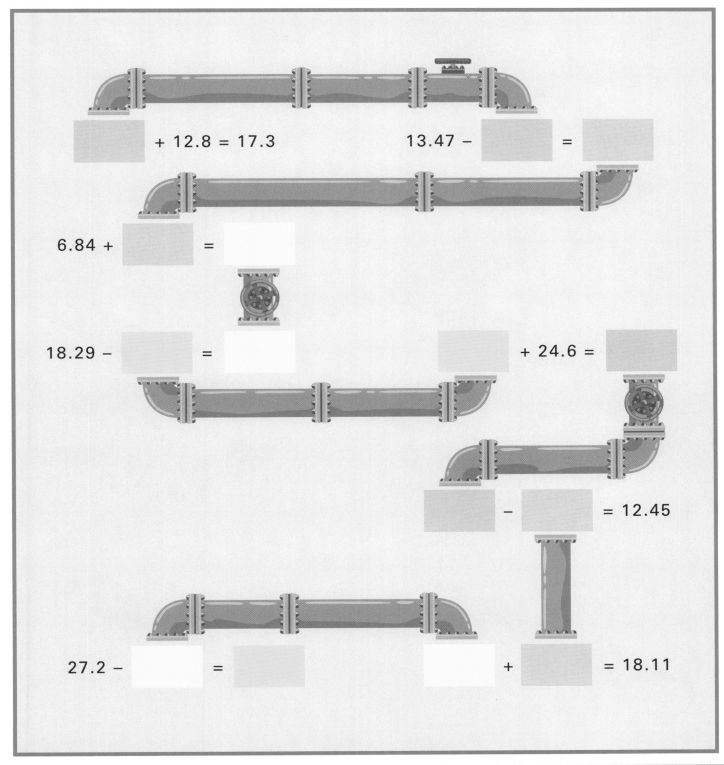

$\square$ + 12.8 = 17.3

13.47 – $\square$ = $\square$

6.84 + $\square$ = $\square$

18.29 – $\square$ = $\square$

$\square$ + 24.6 = $\square$

$\square$ – $\square$ = 12.45

27.2 – $\square$ = $\square$

$\square$ + $\square$ = 18.11

Adding & Subtracting Decimals

Crossing Paths

WRITE the missing numbers.

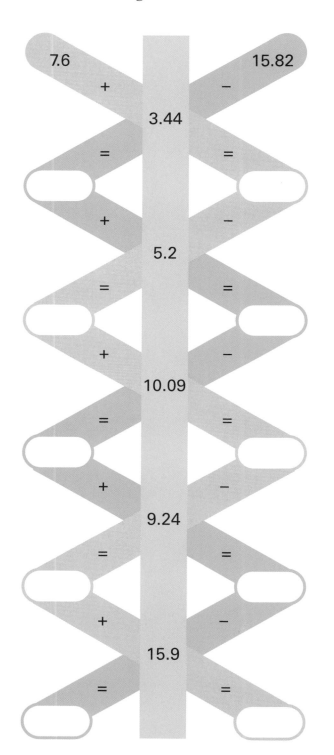

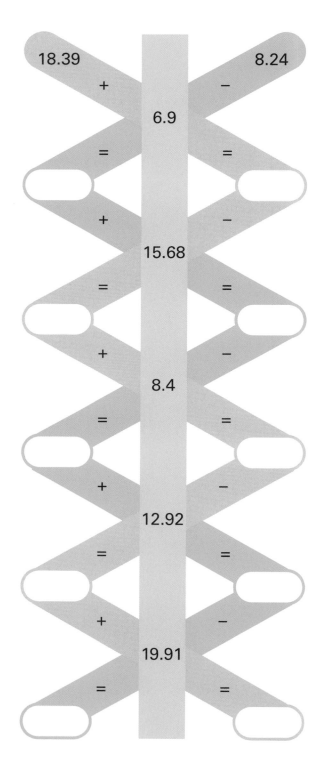

What's the Password?

WRITE the letters that form a part of each word. Then WRITE the letters in order to find the secret password.

1. The first 0.2 of **FOUNDATION** _____

2. The middle $\frac{2}{4}$ of **MOTH** _____

3. The first 0.1 of **BACKPACKED** _____

4. The first $\frac{1}{4}$ of **ALARMING** _____

5. The last $\frac{1}{9}$ of **PRINCIPAL** _____

6. The first 0.2 of **PLAYGROUND** _____

7. The last $\frac{1}{3}$ of **DISMAY** _____

8. The last 0.3 of **VOLUNTEERS** _____

Password:

_____ _____ _____ _____ _____ _____ _____ _____

_____ _____ _____ _____ _____ _____ _____

Number Nudge

For each set of problems, WRITE the missing numbers so that the answers are correct.

0.76	4.25	1.88	5.67

1. $6.29 +$ _____ $=$ 8.17

2. $0.53 +$ _____ $=$ 4.78

3. $2.91 +$ _____ $+$ _____ $=$ 9.34

2.6	5.16	8.43	5.57

4. $14.5 -$ _____ $=$ 6.07

5. $13.19 -$ _____ $=$ 7.62

6. $10.92 -$ _____ $-$ _____ $=$ 3.16

12.87	16.39	14.06	7.82

7. $9.3 +$ _____ $=$ 22.17

8. $6.31 +$ _____ $=$ 14.13

9. $13.75 +$ _____ $-$ _____ $=$ 16.08

Code Ruler

WRITE the letter that matches each measurement to answer the riddle.

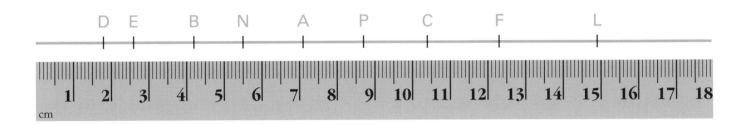

D   E   B   N   A   P   C   F   L

What runs around the yard without moving?

_____   _____   _____   _____   _____   _____ .

7.1 cm   12.3 cm   2.6 cm   5.5 cm   10.4 cm   2.6 cm

Totally Tangled

FIND the measurements that are connected. COLOR the smaller measurement in each pair.

1 centimeter (cm) = 10 millimeters (mm)
1 meter (m) = 100 centimeters
1 kilometer (km) = 1,000 meters

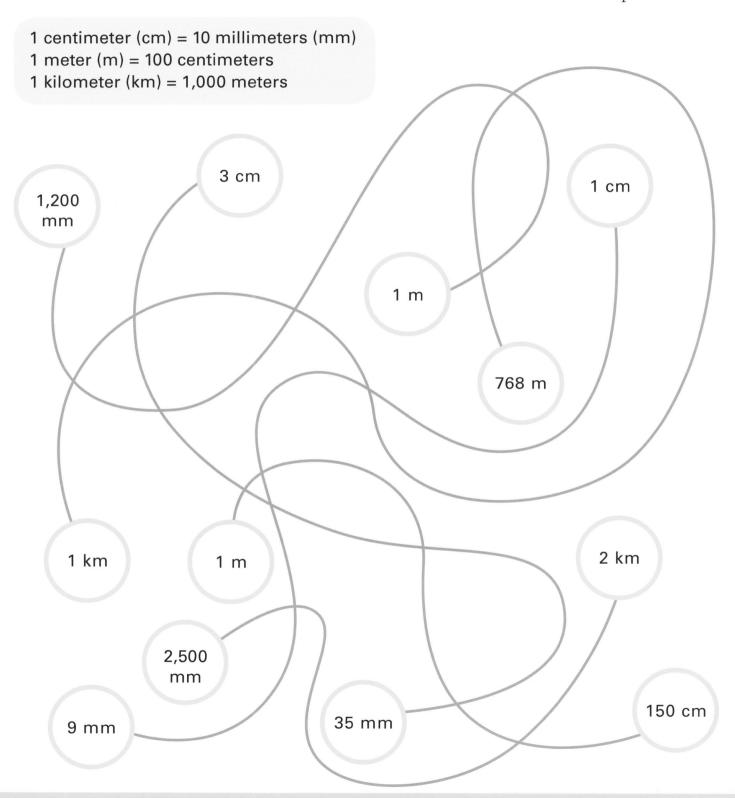

Code Ruler

WRITE the letter that matches each measurement to answer the riddle.

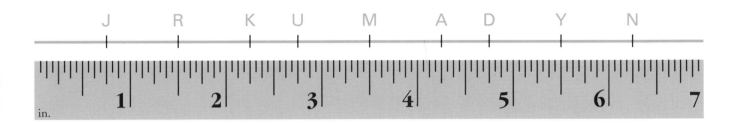

What do you call three feet of garbage?

$\underset{4\frac{1}{4}\text{ in.}}{\underline{A}}$ $\underset{\frac{3}{4}\text{ in.}}{\underline{J}}$ $\underset{2\frac{3}{4}\text{ in.}}{\underline{U}}$ $\underset{6\frac{1}{4}\text{ in.}}{\underline{N}}$ $\underset{2\frac{1}{4}\text{ in.}}{\underline{K}}$ $\underset{5\frac{1}{2}\text{ in.}}{\underline{Y}}$ $\underset{4\frac{1}{4}\text{ in.}}{\underline{A}}$ $\underset{1\frac{1}{2}\text{ in.}}{\underline{R}}$ $\underset{4\frac{3}{4}\text{ in.}}{\underline{D}}$.

Totally Tangled

FIND the measurements that are connected. COLOR the larger measurement in each pair.

1 foot (ft) = 12 inches (in.)
1 yard (yd) = 3 feet
1 mile (mi) = 1,760 yards or 5,280 feet

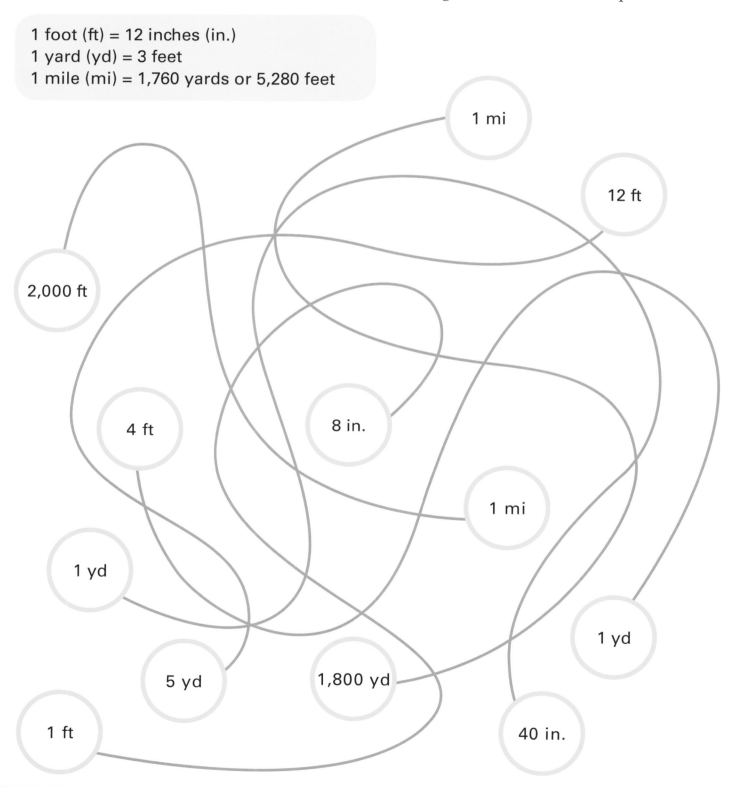

Puzzling Pentominoes

Perimeter is the distance around a two-dimensional shape. Use the pentomino pieces from page 113, and PLACE the pieces to completely fill each shape without overlapping any pieces. Then WRITE the perimeter of each shape. (Save the pentomino pieces to use again later in the workbook.)

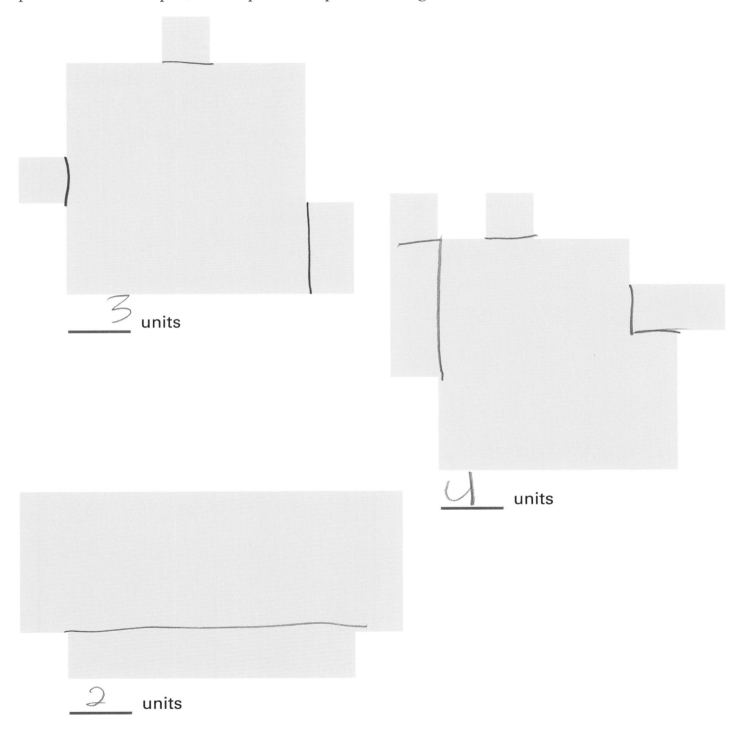

_____3_____ units

_____4_____ units

_____2_____ units

Shape Creator

DRAW three different shapes that all have a perimeter of 12 units.

Using a centimeter ruler, DRAW two different shapes with a perimeter of 20 centimeters.

Puzzling Pentominoes

Area is the size of the surface of a shape, and it is measured in square units. Use the pentomino pieces from page 113, and PLACE the pieces to completely fill each shape without overlapping any pieces. Then WRITE the area of each shape. (Save the pentomino pieces to use again.)

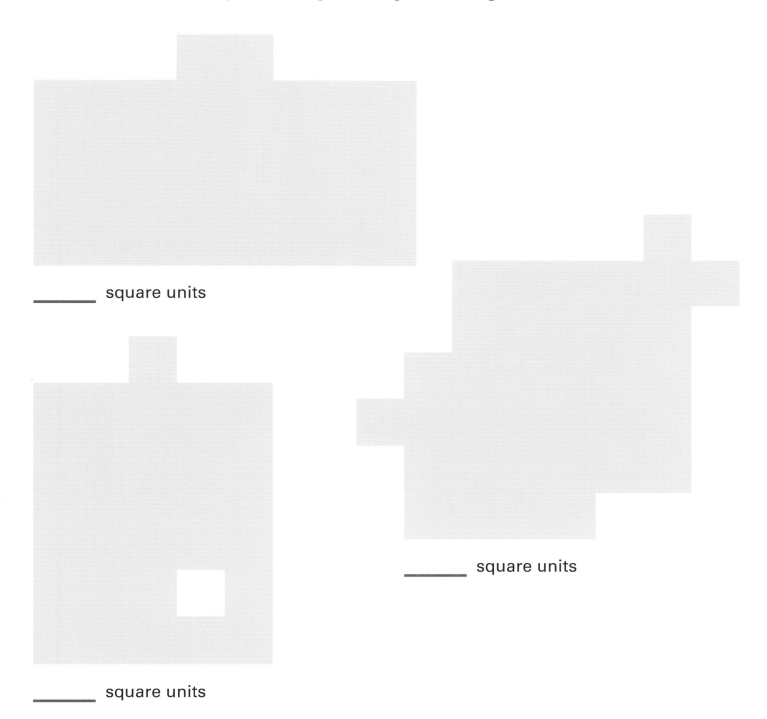

_____ square units

_____ square units

_____ square units

Shape Creator

DRAW three different shapes that all have an area of 10 square units.

Using a centimeter ruler, DRAW two different rectangles with an area of 24 square centimeters.

Code Breaker

SOLVE each problem. WRITE the letter that matches each equivalent measurement to solve the riddle.

1 gram (g) = 1,000 milligrams (mg)	1 kilogram (kg) = 1,000 grams

1 5 g = _____ mg	D	2 2,300 g = _____ kg	L
3 6 kg = _____ g	O	4 600 mg = _____ g	A
5 3,000 mg = _____ g	H	6 6.5 g = _____ mg	U
7 1.5 kg = _____ g	C	8 10,400 g = _____ kg	N
9 0.2 g = _____ mg	Y	10 1,000,000 mg = _____ kg	E

What can you add to a barrel to make it lighter?

___ ___ ___   ___ ___ ___   ___ ___ ___
200   6,000   6,500   1,500   0.6   10.4   0.6   5,000   5,000

___   ___ ___ ___ ___ .
0.6   3   6,000   2.3   1

Totally Tangled

FIND the measurements that are connected. COLOR the smaller measurement in each pair.

1 gram (g) = 1,000 milligrams (mg)   1 kilogram (kg) = 1,000 grams

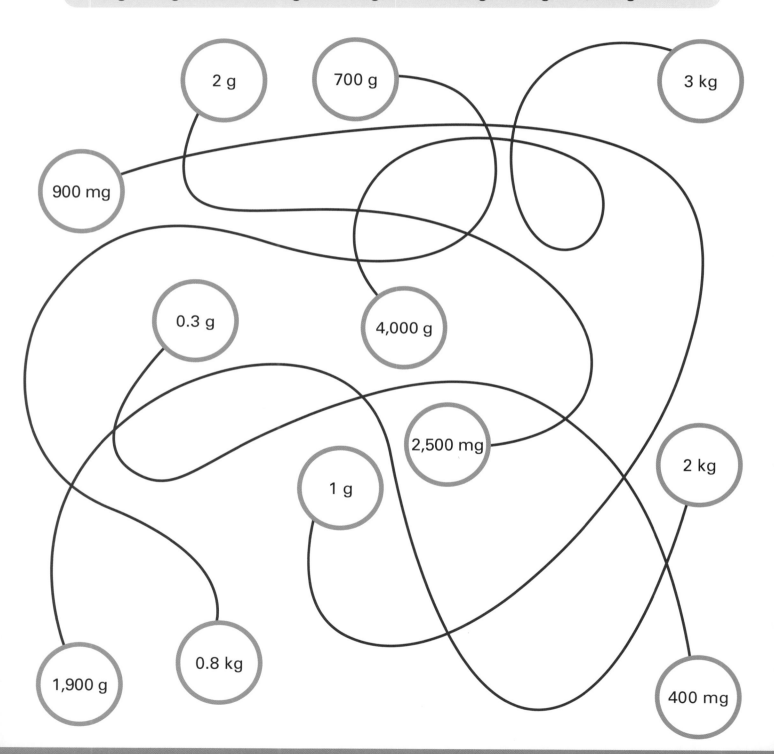

Code Breaker

SOLVE each problem. Use a fraction where necessary. WRITE the letter that matches each equivalent measurement to solve the riddle.

1 pound (lb) = 16 ounces (oz)	1 ton (T) = 2,000 pounds

1 4,000 lb = _____ T	T	2 24 oz = _____ lb	I
3 2 lb = _____ oz	W	4 $\frac{1}{4}$ T = _____ lb	N
5 3 T = _____ lb	Y	6 5,000 lb = _____ T	B
7 32,000 oz = _____ T	O	8 $\frac{3}{4}$ lb = _____ oz	G
9 $\frac{1}{2}$ lb = _____ oz	H	10 $2\frac{3}{4}$ T = _____ lb	E

What weighs more, a ton of rocks or a ton of leaves?

___ ___ ___ ___   ___ ___ ___ ___
2   8   5,500   6,000 $2\frac{1}{2}$ 1   2   8

___ ___ ___ ___ ___
32   5,500 $1\frac{1}{2}$ 12   8

___ ___ ___   ___ ___ ___ .
1   500   5,500   2   1   500

Totally Tangled

FIND the measurements that are connected. COLOR the larger measurement in each pair.

1 pound (lb) = 16 ounces (oz)   1 ton (T) = 2,000 pounds

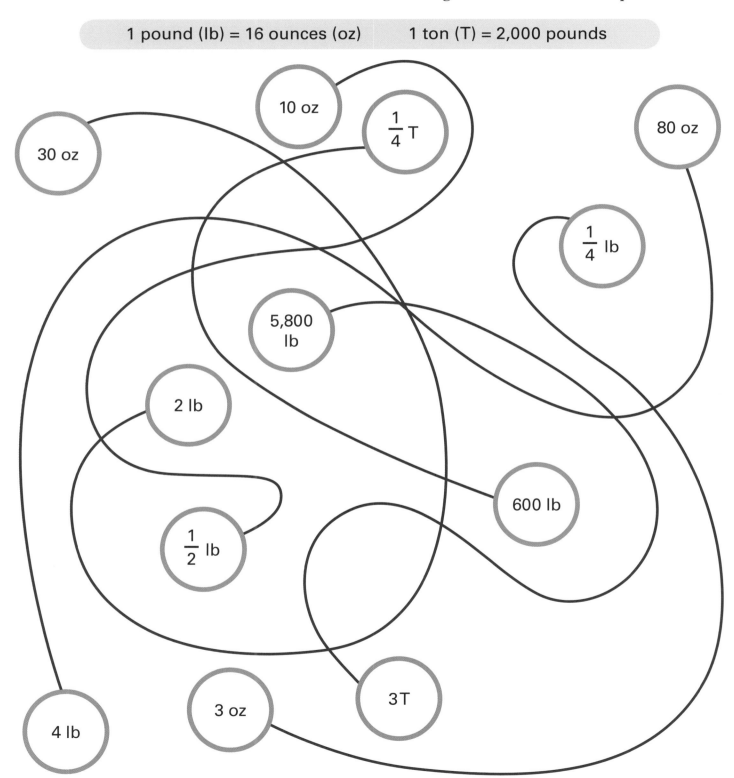

Code Ruler

WRITE the letter that matches each measurement to answer the riddle.

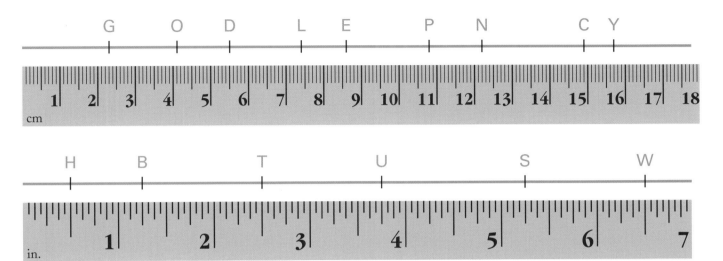

Why did the girl sleep with a ruler?

———— ———— ———— ———— ————
$2\frac{1}{2}$ in. 4.1 cm $5\frac{1}{4}$ in. 8.6 cm 8.6 cm

———— ———— ———— ———— ———— ———— ————
$\frac{1}{2}$ in. 4.1 cm $6\frac{1}{2}$ in. 7.4 cm 4.1 cm 12.2 cm 2.3 cm

———— ———— ———— ———— ———— ———— ———— ————
$5\frac{1}{4}$ in. $\frac{1}{2}$ in. 8.6 cm 14.9 cm 4.1 cm $3\frac{3}{4}$ in. 7.4 cm 5.5 cm

———— ———— ———— ———— ————.
$5\frac{1}{4}$ in. 7.4 cm 8.6 cm 8.6 cm 10.8 cm

Puzzling Pentominoes

Use the pentomino pieces from page 113, and PLACE the pieces to completely fill the shape without overlapping any pieces. Then WRITE the perimeter and area of the shape.

Perimeter: _____ units

Area: _____ square units

Who Am I?

READ the clues, and CIRCLE the mystery shape.

HINT: Cross out any shape that does not match the clues.

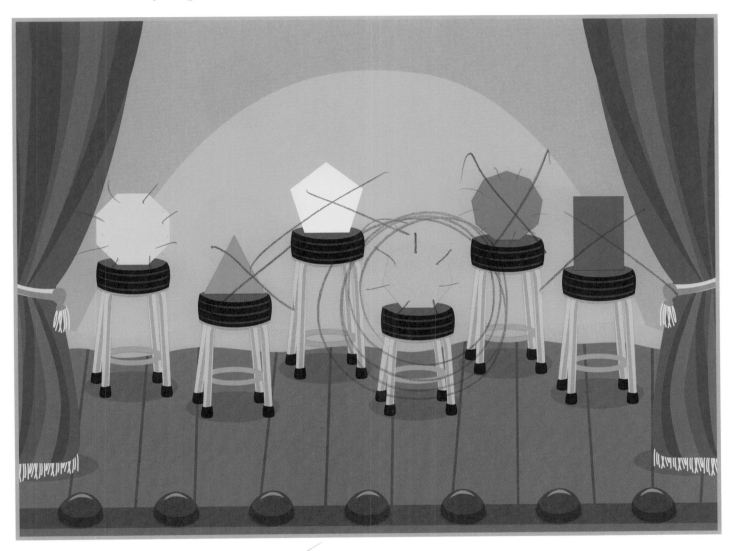

I have fewer than nine sides. ✓

I have no acute angles. ✓

I have more than five sides. ✗

I have seven vertices. ✓

Who am I?

A hexagon

Criss Cross

IDENTIFY each shape, and WRITE the shape names in the puzzle.

ACROSS

2.

3.

6.

7.

DOWN

1.

2.

4.

5.

Fold the Square

Origami is the Japanese art of paper folding, and it often starts with just one simple square. CUT OUT a square piece of paper that is at least six inches by six inches. FOLLOW the steps to make your own paper box.

HINT: Be sure to get a nice crease with each fold.

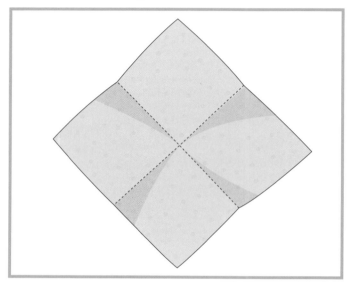

1. Fold the square in half both ways to find a center point.

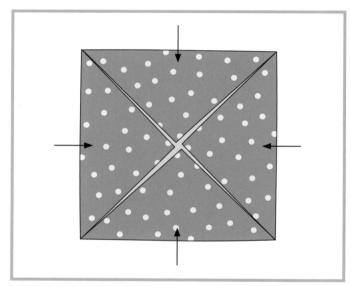

2. Fold the corners into the center, forming a square made from four triangles.

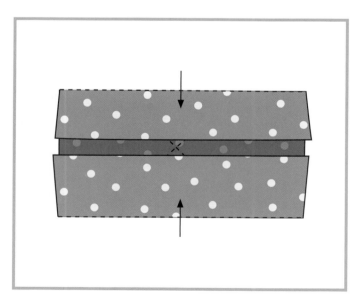

3. Fold two opposite sides into the center.

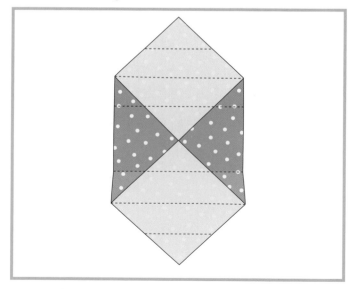

4. Unfold the same two sides until the points are pointing out.

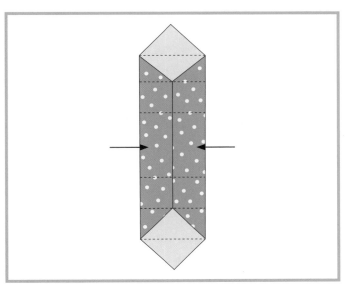

5. Now fold the other two sides into the center.

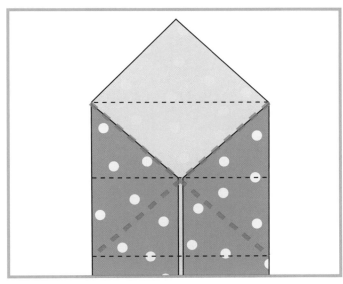

6a. Fold toward you along the angled (red dotted) lines to make creases in the paper.

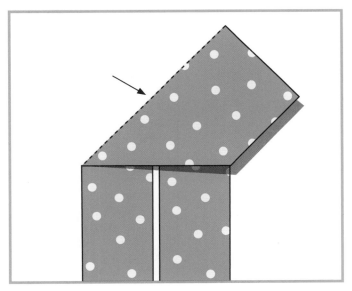

6b. Fold the top part to the right. Then unfold and fold to the left. Creating these creases will make the box easier to fold.

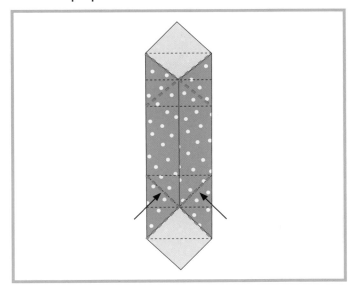

6c. Repeat this on both ends until you've made four creases (red dotted lines).

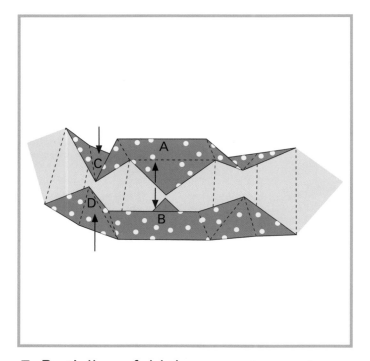

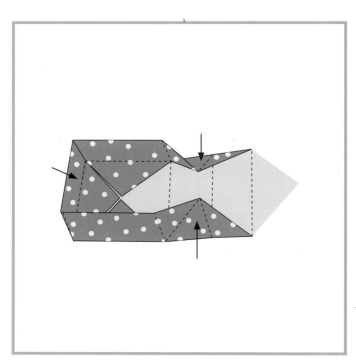

7. Partially unfold the paper to create the left and right sides of the box. Lift up at A and B while pushing in at C and D.

8. Follow the creases of the paper and tuck one end of the long side into the bottom of the box. Repeat on the other side.

Your box is finished and you're ready to store your treats and treasures.

Try following these steps with a larger piece of paper. You can also make a box with a lid if you use two different squares, one slightly larger than the other.

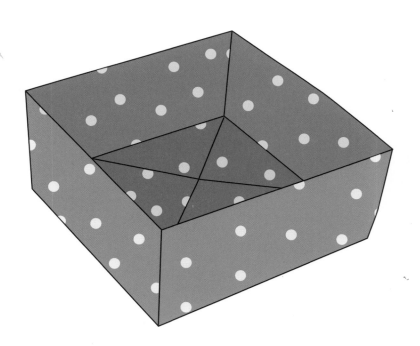

Who Am I?

READ the clues, and CIRCLE the mystery shape.

HINT: Cross out any shape that does not match the clues.

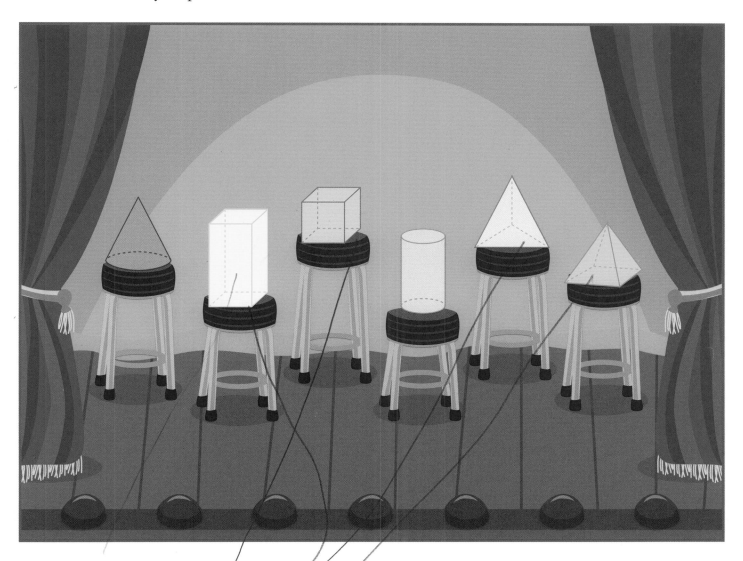

You'll find no round face on me.

I have more than six edges.

I have at least two square faces.

I have four rectangular faces.

Who am I?

trangle   idont know

Criss Cross

IDENTIFY each shape, and WRITE the shape names in the puzzle.

ACROSS

3.

4.

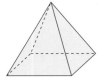

5.

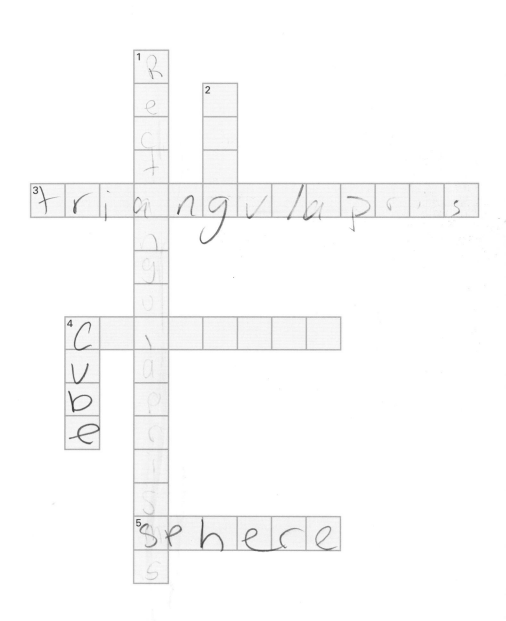

DOWN

1.

2.

4.

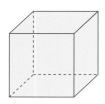

Shape Scavenging

Use these scorecards, and go on a shape scavenger hunt. READ the rules. PLAY the game!

Rules: Two players or teams
1. Pick a location and a time limit for your scavenger hunt. For example, your scavenger hunt can be inside the house for 20 minutes, or each player can pick a different room.
2. When you find something that has the same shape as the shapes on the scorecards, write its name. Shapes that are harder to find earn more points.
3. At the end of the scavenger hunt, add up your points.

The player or team with the most points wins!

| Rectangular prism: 5 points | Cylinder: 8 points | Sphere: 10 points |
| Cube: 12 points | Cone: 15 points | Square pyramid: 20 points |

PLAYER 1

	Items Found	Points
	Total Points	

PLAYER 2

	Items Found	Points
	Total Points	

Shape Builders

CUT OUT each shape on the opposite page. FOLD on the dotted lines, and GLUE the tabs to construct each solid shape. Then WRITE the answers to the questions.

1. What is the name of the blue shape? _____

2. What is the name of the yellow shape? _____

3. Which shape is taller? _____

4. How many faces does the blue shape have? _____

5. How many faces does the yellow shape have? _____

Tricky Tangrams

Use the tangram pieces from page 115, and PLACE the pieces to completely fill each shape without overlapping any pieces. (Save the tangram pieces to use again.)

HINT: Try placing the biggest pieces first.

Tricky Tangrams

Use the tangram pieces from page 115, and PLACE the pieces to completely fill each shape without overlapping any pieces. (Save the tangram pieces to use again.)

HINT: Try placing the biggest pieces first.

Incredible Illusions

A **tessellation** is a repeating pattern of shapes that has no gaps or overlapping shapes. COLOR the rest of the tessellation. Do the rows of shapes look straight or bent?

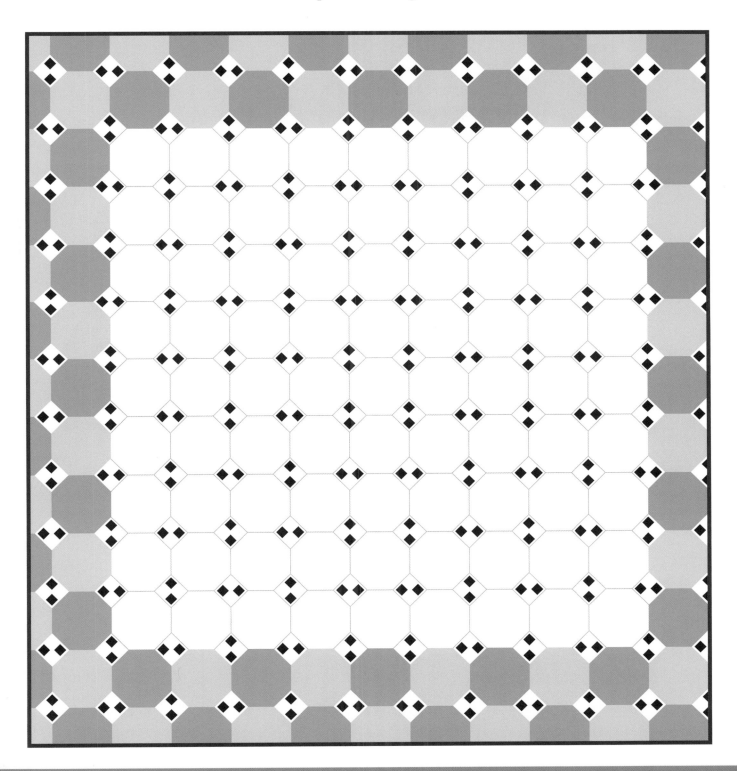

Shape Shifters

Use the pattern block pieces from page 117, and PLACE the pieces to finish the tessellation without overlapping any pieces. (Save the pattern block pieces to use again.)

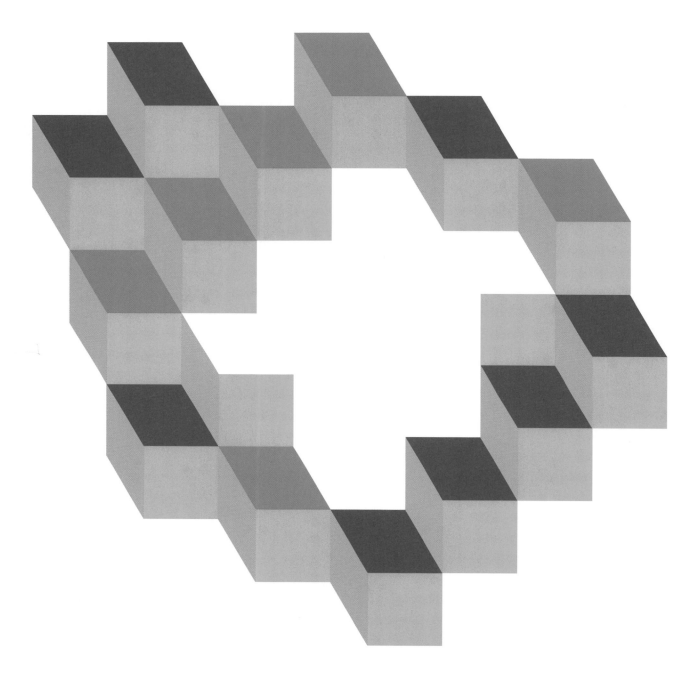

Tricky Tangrams

Use the tangram pieces from page 115, and PLACE the pieces to completely fill each shape without overlapping any pieces.

HINT: Try placing the biggest pieces first.

Shape Shifters

Use the pattern block pieces from page 117, and PLACE the pieces to finish the tessellation without overlapping any pieces.

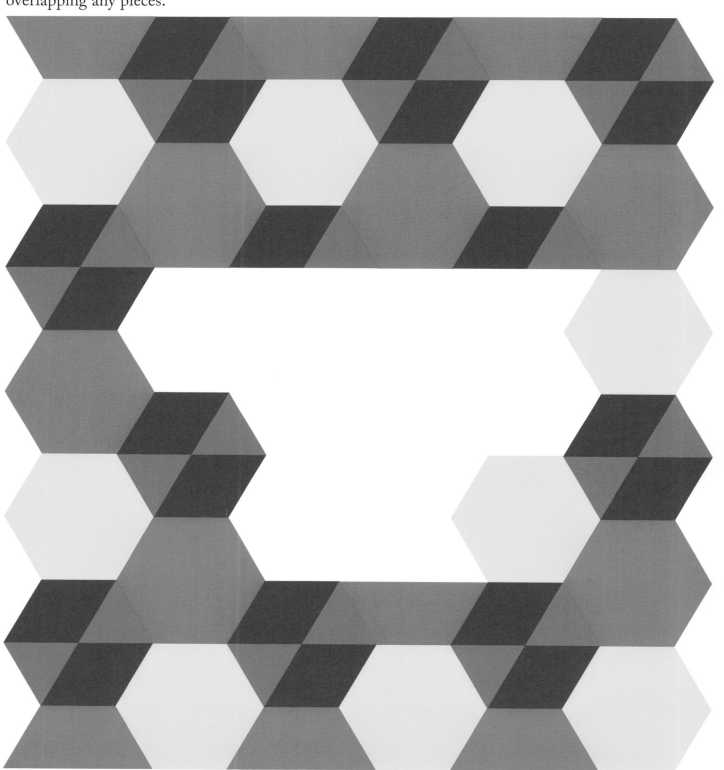

T-Shirt Shop

READ the paragraph, and WRITE the answer.

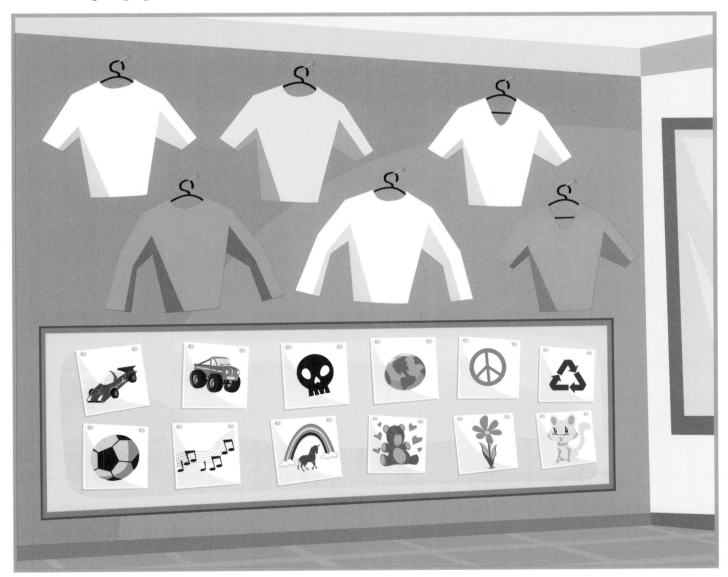

A T-shirt shop can put any design on the T-shirt of your choice. There are 6 different T-shirts and 12 different designs, and you can choose to have your name put on the back or not. How many different T-shirts can you make?

_____ T-shirts

All You Can Eat

READ the paragraph, and WRITE the answer.

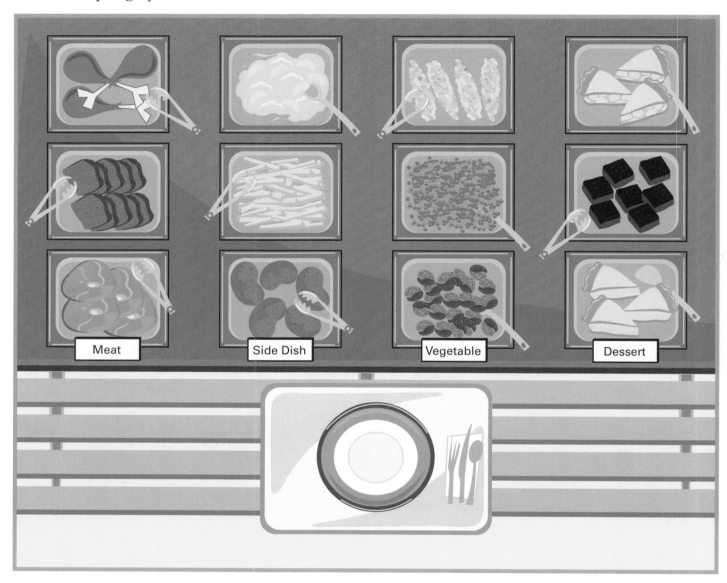

Meat · Side Dish · Vegetable · Dessert

Each time you go to the buffet, your plate should have a choice of one meat, one side dish, one vegetable, and one dessert. How many times can you visit the buffet and get a different plate of food?

_____ times

Bus Ride

READ the clues, and CIRCLE the answer.

The Gallagher sisters always sit together.
Andrew sits next to Alyssa and behind Bill.
Stella likes to sit in back.
Bill always takes the window seat next to Nolan.
Kayley sits in front of Becky.
Dan sits in the aisle seat next to Ella.

Where is Dan in this picture?

In the Neighborhood

WRITE the name of each family on the correct mailbox.

The Green family chose their house for its color.

The Park family is always complaining about the noise coming from the Taguchi house next door.

The Simpsons live across the street from the Green family.

The Taguchis don't like looking out their front window at the Links' lawn flamingoes.

The Meyers live between the Links and the Simpsons.

Colorful Campground

Each tent is a different color. READ the clues, and COLOR each tent red, blue, yellow, green, orange, or purple.

The blue tent is west of the road.

The orange tent is below the lake.

The purple tent is north of the green tent.

The tent farthest south is not orange or purple.

The tent closest to the lake is green.

The red tent has a view of the entire campground.

Secret Location

FIND the points on the map, and WRITE the name of the country at those coordinates. Then UNSCRAMBLE the letters in red to find the secret location on the map.

HINT: Find the first number along the bottom, and the second number along the side. Then find the point where the two lines meet.

1. | 2, 8 | I R E R L A N
2. | 3, 2 | S P A I N
3. | 8, 3 | I T A L Y
4. | 7, 11 | N O R W A Y
5. | 13, 8 | B E L A R U S
6. | 5, 4 | F R A N C E
7. | 7, 7 | G e r m a n y

Secret location: F i n l a n d

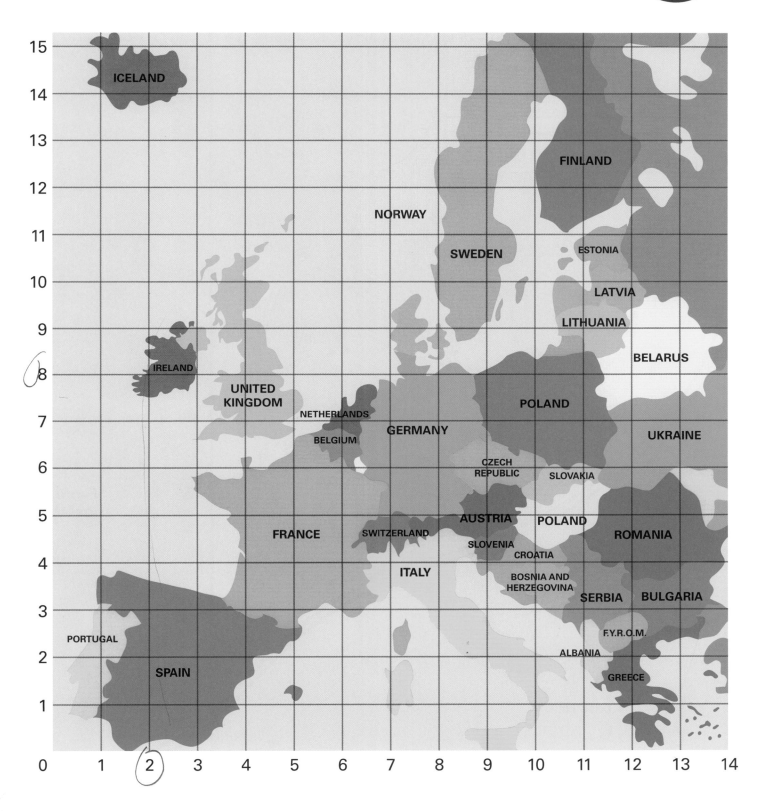

Distant Places

DRAW lines between the four pairs of towns that have a 20-mile stretch of road between them.

HINT: Use the map key to help you.

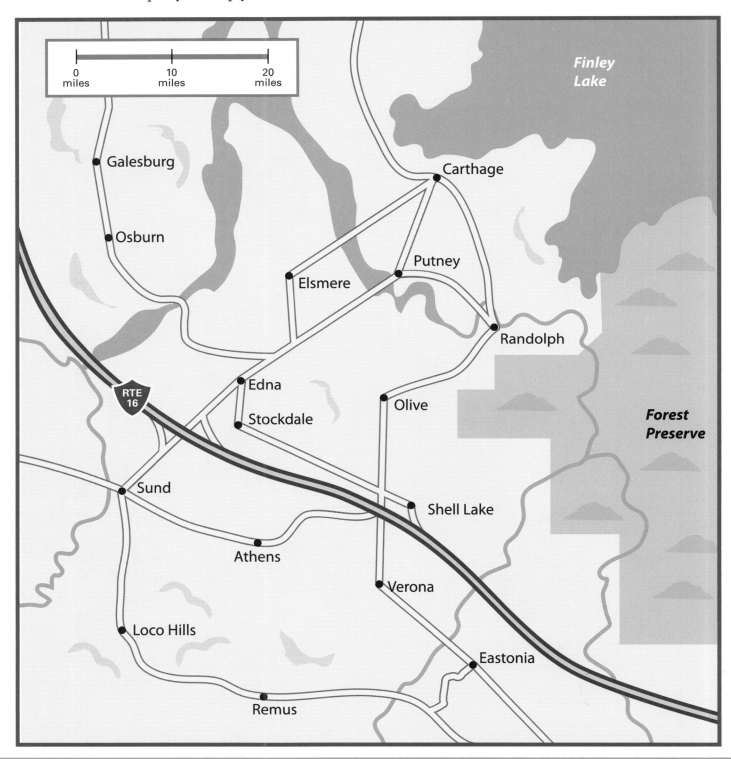

Flip a Coin

Probability is used to describe the chance of something happening. It can be represented by a number from 0 to 1.

Example: The probability that you will grow to be 30 feet tall is 0.

The probability that the sun will rise tomorrow morning is 1.

The probability of getting heads on a coin flip is $\frac{1}{2}$.

To see how probability works, try playing this game. FLIP a coin three times, and WRITE whether you flip heads or tails. SCORE 20 points for turns where you flip either three heads or three tails. SCORE 8 points for turns where you flip heads, heads, tails or tails, tails, heads. SCORE 5 points for any other coin combination. REPEAT this three more times. Then ANSWER the questions.

1	2	3	Score

1	2	3	Score

1	2	3	Score

1	2	3	Score

Total Score: _____

1. What is the chance of scoring 80 points in this game?
   impossible   unlikely   likely   certain

2. What is the chance of scoring 15 points in this game?
   impossible   unlikely   likely   certain

3. What is the chance of scoring at least 20 points in this game?
   impossible   unlikely   likely   certain

Cat's out of the Bag

CUT OUT the pieces from the opposite page, and PUT them in a bag. READ the rules. PLAY the game! Then ANSWER the questions.

HINT: Think about the fraction of each animal to the total number of cards.

Rules: Two players
1. Take turns picking cards out of the bag.
2. Keep picking until all three cats have been found.

The player with the most cats wins!

There are 3 cats, 9 dogs, and 12 mice.

1. What is the probability that you will pull a cat out of the bag on the first turn?

   0     $\frac{1}{8}$ $\frac{1}{4}$ $\frac{1}{2}$ 1

2. What is the probability that you will pull a mouse out of the bag on the first turn?

   0     $\frac{1}{8}$ $\frac{1}{4}$ $\frac{1}{2}$ 1

3. How likely is it that the game will be a tie?

   impossible   unlikely   likely   certain

4. How likely is it that the game will be over after three turns?

   impossible   unlikely   likely   certain

5. How likely is it that three cats will be pulled out of the bag before the game ends?

   impossible   unlikely   likely   certain

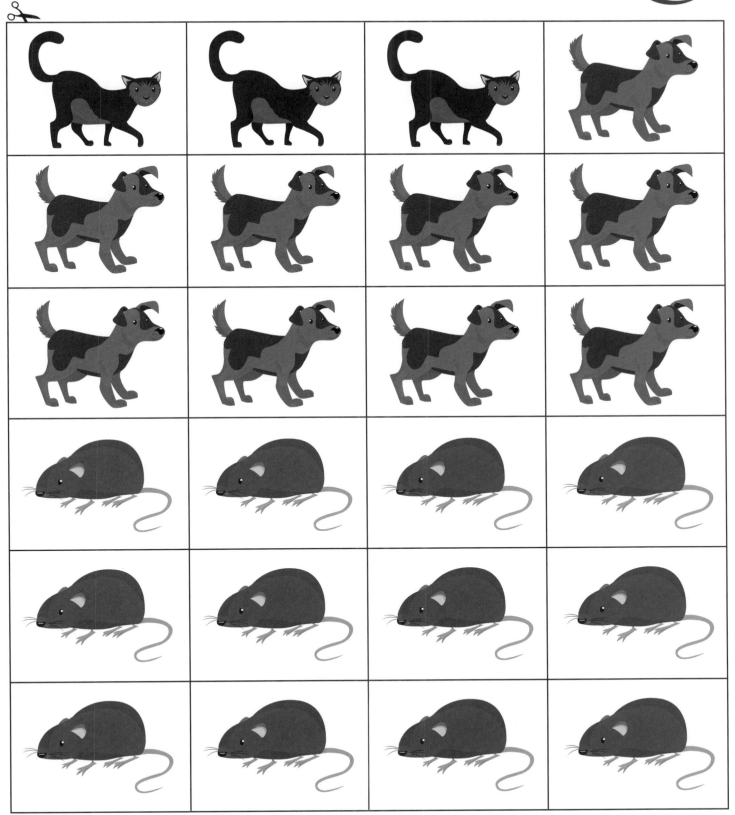

Time Twins

In one day, how many times does a clock have times where all of the digits are the same?

HINT: Don't forget a.m. and p.m.

_2_ times

Rail Race

READ the clues. Then WRITE the number of the train each person took, and CIRCLE the person who got to Washington DC first.

My train departed at 3:50 p.m., exactly 8 minutes late.

Train __41__
₁

I live near Boston. I left the house at 11:00, and it takes me about 45 minutes to get to the train station.

Train __88__
₂

My mom and I missed the 8:24 train in Providence, so we caught the next one.

Train __75__
₃

I always take the train that gets me from Wilmington to Washington in the shortest amount of time.

Train __93__
₄

TRAIN SCHEDULE

	33	75	19	41	68	92
Boston, MA	7:42 a.m.	9:35 a.m.	10:40 a.m.	11:25 a.m.	12:31 p.m.	1:47 p.m.
Providence, RI	8:24 a.m.	10:16 a.m.	11:22 a.m.	12:05 p.m.	1:11 p.m.	2:26 p.m.
New Haven, CT	10:18 a.m.	12:11 p.m.	1:20 p.m.	2:01 p.m.	3:10 p.m.	4:25 p.m.
New York, NY	11:58 a.m.	1:50 p.m.	2:59 p.m.	3:42 p.m.	4:49 p.m.	6:07 p.m.
Trenton, NJ	1:01 p.m.	2:53 p.m.	4:02 p.m.	4:47 p.m.	5:53 p.m.	7:10 p.m.
Wilmington, DE	2:06 p.m.	3:54 p.m.	5:10 p.m.	5:50 p.m.	7:03 p.m.	8:18 p.m.
Baltimore, MD	3:14 p.m.	5:02 p.m.	6:14 p.m.	6:58 p.m.	8:09 p.m.	9:24 p.m.
Washington DC	4:30 p.m.	6:13 p.m.	7:21 p.m.	8:10 p.m.	9:19 p.m.	10:38 p.m.

Holding Hands

In one day, how many times do the hour and minute hands cross each other on a clock? WRITE the answer.

HINT: After 12:00, the first time that the clock hands cross each other is around 1:06. Think about what times the clock hands cross, and draw them on the clock to help you count. Try using a watch if you get stuck.

_____ times

Pocket Change

DRAW three straight lines to create six different money sets of equal value.

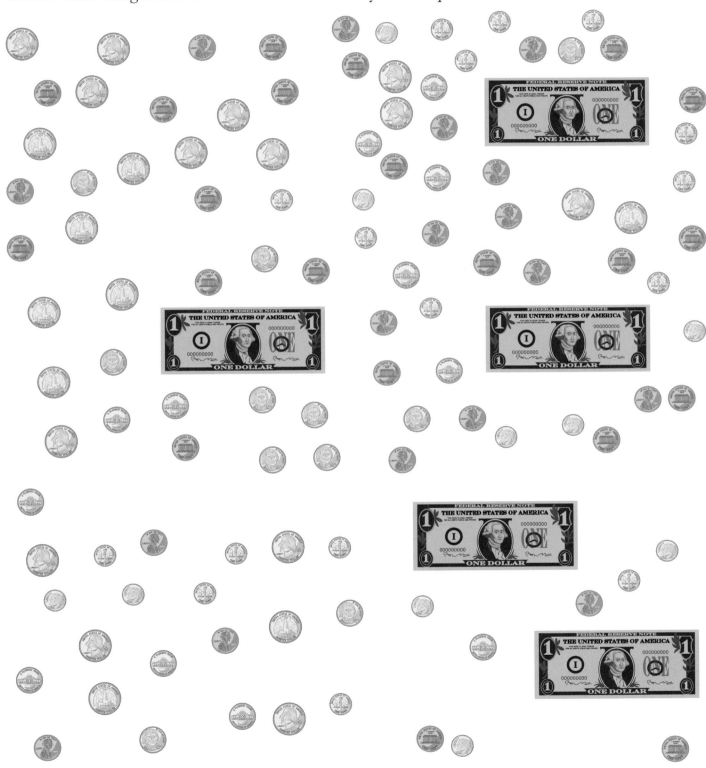

What's in My Hand?

READ the clues, and WRITE how many of each coin and bill are hidden in the hand.

I'm holding six paper bills and nine coins.

The money in my hand totals $48.89.

My coins total less than one dollar.

I don't have any 10-dollar bills.

What's in my hand?

1. _____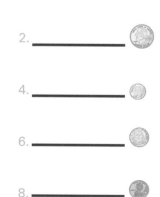

2. _____

3. _____

4. _____

5. _____

6. _____

7. _____

8. _____

Stamp Collector

DRAW lines to mark where you would tear the sheet of stamps to create three sheets of stamps of equal value.

HINT: The three sets do not need to contain the same number of stamps.

Big Spender

Rachel and Gabby went to a Sara Starlight concert together, and Rachel spent twice as much as Gabby.
DRAW a circle around the things that Rachel bought and a square around the things that Gabby bought.

$25.00

$25.00

$35.00

$19.00

$2.75

$6.50

Awesome Avatars

READ the paragraph, and WRITE the answer.

A new video game lets you design your character before you start playing. First you can choose to be a boy or a girl. Then you get a choice of four different skin colors, three different hairstyles, and five different outfits. How many different characters can be made from these choices?

_____ characters

Big Spender

Ethan and Travis went to the store together, and Ethan spent twice as much as Travis. DRAW a circle around the things that Ethan bought and a square around the things that Travis bought.

$46.87

$14.36

$24.57

$18.05

$25.99

$21.33

$4.50

$79.26

Fraction and Decimal Cards

CUT OUT the 24 cards.

These cards are for use with page 45. Use either the fraction side or the decimal side. (The two sides are not equivalent.)

$\frac{1}{2}$	$\frac{1}{3}$	$\frac{2}{3}$	$\frac{1}{4}$
$\frac{3}{4}$	$\frac{1}{5}$	$\frac{2}{5}$	$\frac{3}{5}$
$\frac{4}{5}$	$\frac{1}{6}$	$\frac{5}{6}$	$\frac{1}{7}$
$\frac{2}{7}$	$\frac{3}{7}$	$\frac{4}{7}$	$\frac{5}{7}$
$\frac{6}{7}$	$\frac{1}{8}$	$\frac{3}{8}$	$\frac{5}{8}$
$\frac{7}{8}$	$\frac{1}{9}$	$\frac{4}{9}$	$\frac{8}{9}$

0.1	0.2	0.3	0.4
0.5	0.6	0.7	0.8
0.9	0.16	0.23	0.25
0.33	0.38	0.42	0.48
0.55	0.61	0.67	0.73
0.74	0.87	0.91	0.99

Pentominoes

CUT OUT the 13 pentomino pieces. (Cut along the black lines only.)

These pentomino pieces are for use with pages 61, 63, and 70.

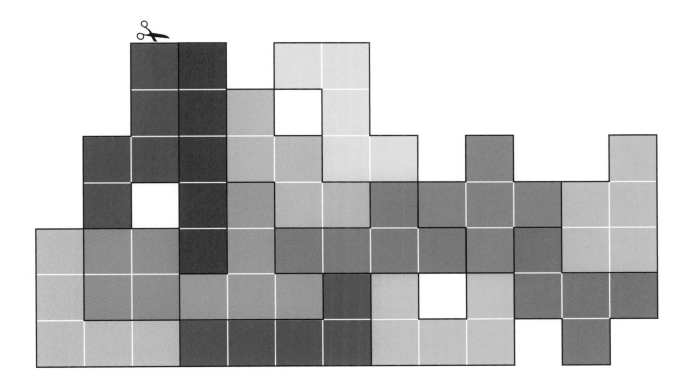

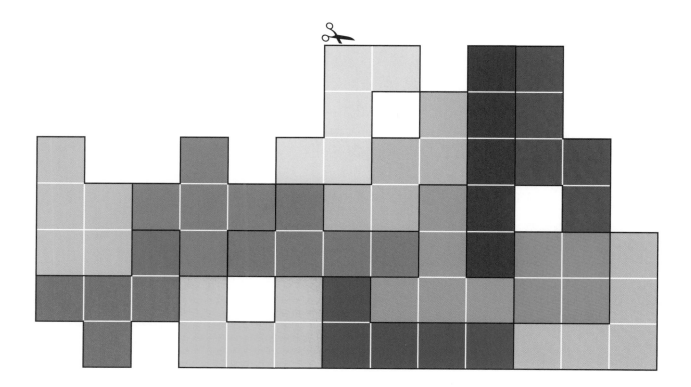

Tangrams

CUT OUT the seven tangram pieces.

These tangram pieces are for use with pages 83, 84, and 87.

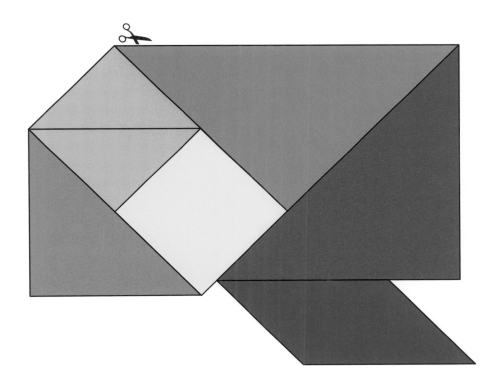

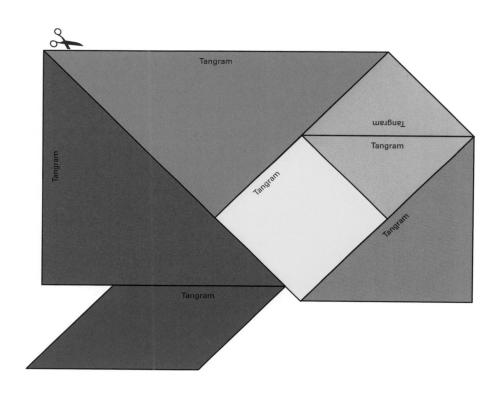

Pattern Blocks

CUT OUT the 31 pattern block pieces.

These pattern block pieces are for use with pages 86 and 88.

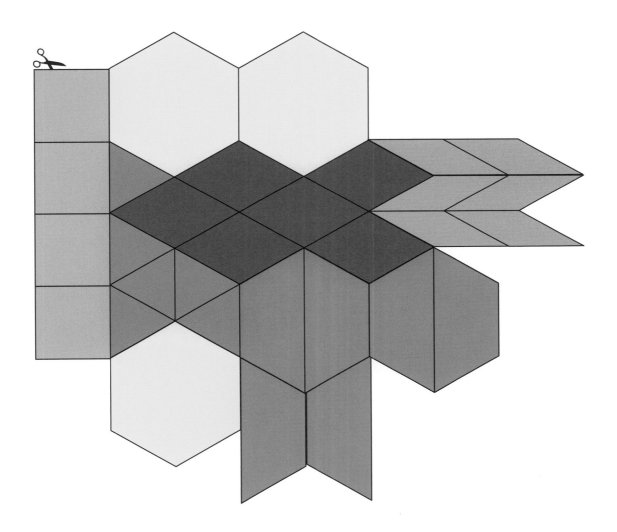

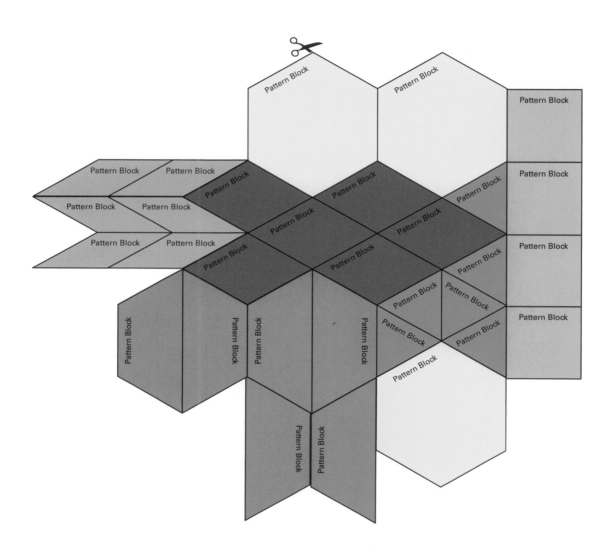

Page 3
1. 84,165 2. 4,672,244
3. 961,723 4. 29,811
5. 115,736 6. 2,082,641
7. 505,692 8. 3,937,260

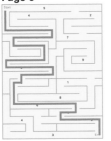

Pages 4–5

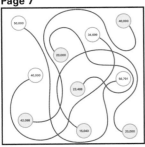

Page 6

7,138,605

Page 7

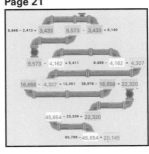

Page 8
1. 228,864 2. 341,156
3. 275,319 4. 384,620
5. 382,495 6. 392,382
7. 232,981 8. 337,236
9. 246,518

Page 9

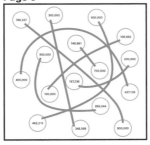

Page 10
1. 5,237,564 2. 5,418,163
3. 5,908,752 4. 6,692,556
5. 6,694,204 6. 6,563,827
7. 5,879,215 8. 5,418,921
9. 5,826,138

Page 11
1. 27, 49, 80, 115
2. 0, 35, 66, 112
3. 106, 165, 240, 294

Page 12

1,839,617

Page 13
1. 21, 33, 45, 63
2. 2, 3, 6, 9
3. 4, 8, 14, 18

Page 14
6,818,567

Page 15
1. 4, 5   2. 2, 7   3. 5, 3

Page 16

two million, six hundred
forty-nine thousand, five
hundred thirty-eight

Page 17

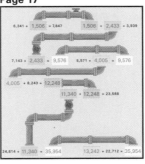

Pages 18–19

Page 20
1. 76,977 2. 85,543
3. 51,349 4. 59,868
5. 99,758 6. 62,372
7. 40,703 8. 71,230

Page 21

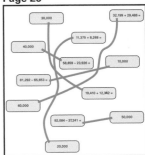

Pages 22–23

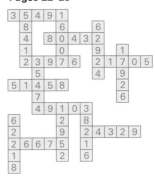

Page 24
1. 22,311 2. 76,213
3. 55,224 4. 71,018
5. 49,876 6. 18,236
7. 61,086 8. 34,972

Page 25

Page 26

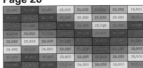

Page 27

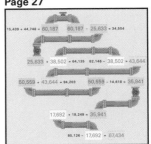

Answers

Page 28

85,000	–	53,218	=	31,782
–		–		–
72,883	–	46,914	=	25,969
=		=		=
12,117	–	6,304	=	5,813

Page 29

1. 30 2. 16 3. 36
4. 25 5. 24 6. 60
7. 7 8. 72 9. 48
10. 18 11. 45 12. 40

WHEREVER YOU PUT HIM.

Page 30

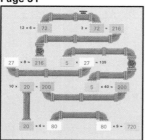

	4	7
1	4	7
2	8	14

	5	6
1	5	6
3	15	18

	2	7
9	18	63
10	20	70

	5	8
2	10	16
6	30	48

	3	8
7	21	56
9	27	72

	5	6
7	35	42
9	45	54

Page 31

12 × 6 = 72
3 × 72 = 216
27 × 8 = 216
5 × 27 = 135
10 × 20 = 200
5 × 40 = 200
20 × 4 = 80
80 × 9 = 720

Pages 32–33

1	0	3	6				
2			8		4		
7			5	6	0	3	2
2	7	3			8		
	5		3	2	0	8	
	1		5			3	
1	4	4	0	5		3	
6			7			2	
5				2	8	8	0
2	5	5	3	2			
				6			
	3	7	4	8	8		

Page 34

13	×	11	=	143
×		×		×
6	×	14	=	84
=		=		=
78	×	154	=	12,012

Page 35

1. 7 2. 8 3. 10
4. 1 5. 4 6. 2
7. 5 8. 11 9. 3
10. 12 11. 6 12. 9

HE CAUGHT A LOT OF FLIES.

Page 36

1. 9, 2, 6, 8
2. 8, 3, 1, 7
3. 16, 4, 8, 10

Page 37

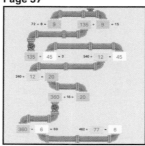

72 ÷ 8 = 9
135 ÷ 9 = 15
135 ÷ 45 = 3
540 ÷ 12 = 45
240 ÷ 12 = 20
360 ÷ 18 = 20
360 ÷ 6 = 60
462 ÷ 77 = 6

Pages 38–39

1	3	2		4	1	8
3	4	1		1	0	5
		0		9		
2	1				1	2
2	5	3		4	6	0
3	0				2	9
		3		5		
2	9	1		3	8	5
1	2	0		2	4	3

Page 40

972	÷	36	=	27
÷		÷		÷
54	÷	6	=	9
=		=		=
18	÷	6	=	3

Page 41

24
112
8
14
192
16
12
224
28
8
336
84
4
672
32
21
128
56
975
15
65
840
8
105
520
4
130
420
7
60
910
91
10
5,460

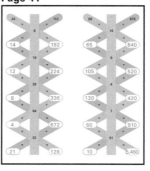

Page 42

1. 20, 180, 372, 724
2. 144, 180, 468, 612
3. 5,984, 6,664, 7,208, 8,024

Page 43

1. SU 2. M 3. ME
4. R 5. VAC 6. A
7. TI 8. ON

SUMMER VACATION

Page 44

0.28
0.8
0.1
0.59
0.43
0.3
0.81
0.7

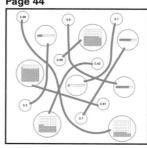

Page 45

Have someone check your answers.

Page 46

1.2
1.23
0.12
0.7
0.3
0.66
0.58
0.31
0.99
0.64
2.1
0.2

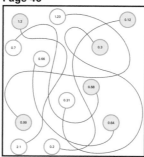

Page 47

1. $\frac{3}{4}$ 2. $\frac{5}{6}$ 3. $\frac{3}{3}$
4. $\frac{5}{9}$ 5. $\frac{7}{8}$ 6. $\frac{2}{5}$
7. $\frac{2}{3}$ 8. $\frac{5}{7}$ 9. $\frac{5}{8}$

THE LIBRARY.

Page 48

1. $\frac{2}{7}$, $\frac{4}{7}$, $\frac{6}{7}$, $\frac{7}{7}$
2. $\frac{3}{9}$, $\frac{5}{9}$, $\frac{6}{9}$, $\frac{8}{9}$
3. $\frac{7}{10}$, $\frac{8}{10}$, $\frac{10}{10}$, $\frac{12}{10}$

Page 49

1. $\frac{2}{5}$ 2. $\frac{7}{12}$ 3. $\frac{3}{8}$
4. $\frac{1}{3}$ 5. $\frac{3}{6}$ 6. $\frac{1}{7}$
7. $\frac{1}{4}$ 8. $\frac{3}{5}$ 9. $\frac{2}{9}$
10. $\frac{2}{11}$ 11. $\frac{5}{6}$ 12. $\frac{3}{10}$

THEY USE THEIR SHELL PHONES.

Page 50

1. $\frac{1}{6}$, $\frac{2}{6}$, $\frac{3}{6}$, $\frac{4}{6}$,
2. $\frac{1}{8}$, $\frac{3}{8}$, $\frac{5}{8}$, $\frac{6}{8}$,
3. $\frac{1}{12}$, $\frac{4}{12}$, $\frac{6}{12}$, $\frac{8}{12}$,

Page 51

1. 9.7 2. 3.5
3. 3.81 4. 8.47
5. 8.02 6. 10.19
7. 6.1 8. 5.6
9. 4.43 10. 1.84
11. 2.93

TIME TO GET A NEW TABLE.

Page 52

1. 2.1, 2.5, 3.4, 4.3
2. 5.69, 6.2, 9.75, 12.18
3. 8.71, 12.59, 16.91, 26.47

Page 53

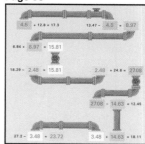

4.5 + 12.8 = 17.3
13.47 - 4.5 = 8.97
6.84 + 8.97 = 15.81
18.29 - 2.48 = 15.81
2.48 + 24.6 = 27.08
27.08 - 14.63 = 12.45
27.2 - 3.48 = 23.72
3.48 + 14.63 = 18.11

Answers

Page 54

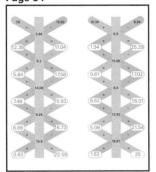

Page 55
1. FO   2. OT   3. B
4. AL   5. L   6. PL
7. AY   8. ERS
FOOTBALL PLAYERS

Page 56
1. 1.88   2. 4.25
3. 5.67, 0.76 4. 8.43
5. 5.57   6. 5.16, 2.6
7. 12.87   8. 7.82
9. 16.39, 14.06

Page 57
A FENCE.

Page 58
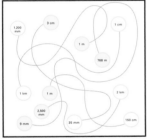

Page 59
A JUNKYARD.

Page 60
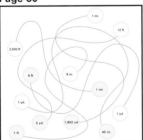

Page 61
Suggestion:

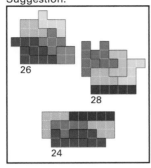

26
28
24

Page 62
Have someone check your answers.

Page 63
Suggestion:

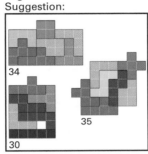

34
35
30

Page 64
Have someone check your answers.

Page 65
1. 5,000 2. 2.3   3. 6,000
4. 0.6   5. 3   6. 6,500
7. 1,500 8. 10.4 9. 200
10. 1
YOU CAN ADD A HOLE.

Page 66
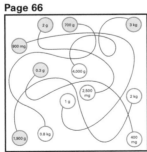

Page 67
1. 2   2. $1\frac{1}{2}$ 3. 32
4. 500   5. 6,000 6. $2\frac{1}{2}$
7. 1   8. 12   9. 8
10. 5,500
THEY BOTH WEIGH ONE TON.

Page 68
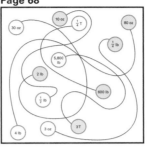

Page 69
TO SEE HOW LONG SHE COULD SLEEP.

Page 70
Suggestion:

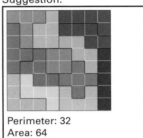

Perimeter: 32
Area: 64

Page 71
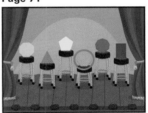

Pages 72–73

Pages 74–76
Have someone check your answers.

Page 77
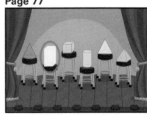

Page 78

Page 79
Have someone check your answers.

Page 80
1. cube
2. square pyramid
3. square pyramid
4. 6
5. 5

Page 83
Suggestion:

Page 84
Suggestion:

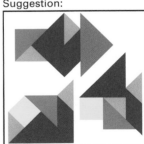

Answers

Page 85

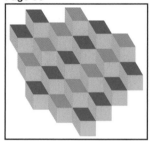

Page 86

Page 87
Suggestion:

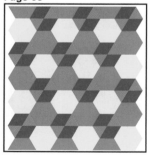

Page 88

Page 89
144

Page 90
81

Page 91

Page 92

Page 93

Pages 94–95
1. IRELAND
2. SPAIN
3. ITALY
4. NORWAY
5. BELARUS
6. FRANCE
7. GERMANY
FINLAND

Page 96

Page 97
1. unlikely
2. impossible
3. certain

Page 98
1. $\frac{1}{8}$ 2. $\frac{1}{2}$
3. impossible
4. unlikely
5. certain

Page 101
12 (1:11, 2:22, 3:33, 4:44, 5:55, 11:11, both a.m. and p.m.)

Pages 102–103
1. 41   2. 68
3. 75   4. 19

Page 104
22 (Clock hands pass each other around the times 1:06, 2:11, 3:17, 4:22, 5:27, 6:33, 7:38, 8:43, 9:49, 10:55, and 12:00 twice per day.)

Page 105

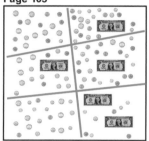

Page 106
1. 2   2. 3
3. 0   4. 0
5. 1   6. 2
7. 3   8. 4

Page 107

Page 108

Page 109
120

Page 110